OUR DAD THE HERO

"...Go with me on my journey..."

By

ROBERT GOSNEY

COLONEL, U.S. ARMY (RETIRED)

ZOOKBOOKS

5923 Denman's Loop, Belton, TX 76513

Published by ZookBooks
Edited by Aaron M. Zook, Jr., www.zookbooks.org,
email: zookaaron@yahoo.com

Printed in the United States of America
ZookBooks, 5923 Denmans Loop, Belton, TX 76513

Library of Congress Control Number: 2015959903
Zook, Aaron
Our Dad, The Hero

ISBN: 978-0-692-59618-0

Cover Photo by James Dietz
Cover and Interior Design by 5 Lakes Design

Disclaimer
This book is designed to document factual accounts of the author's life as recalled through personal stories to inspire readers to discover the value of their own life adventures. Personal opinions expressed in this publication are those of the author and not opinions of the editor, publisher, or any other organization, governmental or civilian.

The author and ZookBooks, and any of their family members shall have neither liability nor responsibility to any person or entity with respect to any loss, damage, injury, or incident alleged to have been caused, directly or indirectly, by the information contained in this book.

Dedication

I dedicate this book to my wonderful wife of sixty-five years, Jean.

Table of Contents

INTRODUCTION

I live life the old-fashioned way—one crisis at a time. Once my brother Bill had me reach into a hole in the ground and yank out a skunk. We both got sprayed and reeked until my Granny swung into action with a home remedy. Another time, I swatted a bee's nest out of a barn's rafters, then raced from the swarm of angry insects to Granny's house. She called our doctor. He yanked out twenty-seven stingers and my face swelled until my eyes couldn't open. Later, as a first-time teen pilot of a two-seater aircraft, my heart pounded during takeoff when the landing gear cleared the dirt runway's bordering trees by inches.

Close calls were in my DNA. They kept coming. In Vietnam, as a helicopter pilot, I attacked the enemy, braving their return fire. In one incident, they blew my chopper's tail off. On my second time to Vietnam, I commanded an Air Mobile Aviation unit in jungle firefights in Vietnam and in Cambodia. Every day marked another penetration into enemy territory. Thrust into hair-raising situations, life and death decisions, and unknown risks, I mastered my emotions and conquered my fears because I crave adventure. Maybe my thirst for action came from earlier generations in my heritage.

My relatives were pioneers, taming their land; hardy people—good stock. My Grandmother escaped renegade Indians in Texas by her quick reaction to hide in a pecan tree's branches. The Indians she evaded killed most of a rancher's family further down the river, leaving the rancher's little daughter as the sole survivor. Granny's dad, Great-Grandfather Miller, led a posse, tracked and located the murderers, and battled them to their deaths before they massacred other victims.

Decades after this encounter, I entered the world in a three-room cabin in 1931, in Junction, Texas. I had an older and a younger brother: Bill, born in 1930 and Gary, born in 1944. We could be a handful. Knocking bee nests down wasn't our only specialty. However, both Mom and Dad corralled us and instilled rock-solid values and a strong work ethic into our lives.

I played football, following my brother Bill's footsteps, in high school (halfback and lineman) and in college (freshman - center, then guard and linebacker the remaining years). I lettered four years in high school and at Texas A&M with no major injuries.

Before I entered college, I struck out on a new adventure. I eloped with my lovely wife, then Jean Hefner, who I met in Handley, Texas. Jean was a cheerleader and I was a football player—an excellent match. We got married in 1950 right before I started college. Marrying Jean was one of the best decisions of my life.

I chose the great Texas A&M University for my college. I wanted tradition and an excellent education for future advancement. The A&M Corps of Cadets was a perfect fit because they allowed married cadets in the Corps and personified the traditions of the University. In addition to the Corps of Cadets' demands, I supported my wife by part-time work, played football, and kept my grades high. Jean worked to help pay the bills and to ensure we had money left over each month. To keep expenses low, sometimes we commandeered steaks from the football player's meals for our wives, fired shotguns off to celebrate the 4th of July, and ate delicious food like frogs legs and baloney delights. Life in college was a hoot.

Graduation from Texas A&M University opened the door to a commission in the United States Army in 1954. Beginning as an Artillery officer, I went to Germany, worked for several larger-than-life bosses, breezed through several military schools, and became a helicopter pilot.

In 1961, disaster struck as Hurricane Carla wiped several towns off the map in Texas. I led a search and rescue program for the Army to relieve misery, privation, and lack of shelter issues to help Texans get back on their feet.

The Army selected me for two tours in the treacherous world of combat in Vietnam in 1966-67 and 1969-70. But when my brother Gary arrived in Vietnam during my last tour, I linked up with him and taught him Vietnam survival skills in the 25th Infantry Division headquarters.

Years after my Vietnam tours, Army assignment managers talked me into a tour in Liberia, Africa, where the world went to hell-in-a-handbasket in one day—twice. The 1979 Rice Riots and the 1980 coup sent shivers through the Liberian nation, but the United States Military Mission established control in the capital city of Monrovia both times and rescued a nation from the brink of disaster. In both of these events, I played a key role. I took control of the Liberian Army in the Rice Riots and made critical governmental decisions with the 1980 coup leaders within the first twenty-four hours after they seized power. Faced with deadly decisions, gun-toting soldiers, and a breakdown of law and order, I helped the Liberian leadership keep their country from disintegrating.

Changing venues from world events in Africa didn't slow the pace of my life's unusual events. I spent my last four years in the military working with the civilian community around Fort Hood and life hummed while I met the needs of the Commander and the surrounding towns.

After retiring from the military, life continued to throw

challenges in my direction. I jumped at each one as an opportunity, capitalizing on seeking and finding the best avenue to support my loved ones, my community and those with whom I served this nation.

But in the end, this story is one about a boy who loved adventure, his family, the American people, and the value of doing the right thing in any situation.

Dig into my adventures, unearth the golden nuggets hidden within, and capture the time-honored principles reflected in the following pages. Maybe you'll be inspired to record your own life-adventure stories.

CHAPTER 1
How I Learned to Fly

Passion for flying burned deep within even as a boy. An older man struck a match, the beginning of an internal flame, by strapping me inside an airplane and taking me for a gut-wrenching flight. I never forgot that event. About forty years after the first plane flew at Kitty Hawk in 1903, aviation blossomed into a new transportation industry. Aircraft were game-changing machines in World War I and World War II combat. My first taste of aerial freedom came from a prior military pilot in a Stearman Biplane aircraft, an aircraft that had been in development and service since the 1920s.

Flying in a Stearman Airplane

The one time I ever got sick flying was when a friend, Roy Cashine, flew me around in the Stearman. His loops and fancy military maneuvers kept my knuckles white from hanging on.

Up a decent rise at Forest Hill airport, there was a little grass runway with a small hanger. An old man out there ran the operation. But Roy, a World War II Navy pilot, had a Stearman plane parked at the airport and asked me one day if I wanted a ride.

Heck, I wasn't even a freshman in high school. I was about twelve years old.

"You bet I'd go flying," I said.

The only time he kept the plane level was when we took off and when we landed. That Stearman was acrobatic to the core. Roy did spins, inverted maneuvers, and other Navy combat evasive techniques.

I lost my lunch all over the airplane. Good grief, did I ever get my guts twisted in a knot. I even coughed up the nuts I ate when I was five years old...not really. But I emptied out everything I had.

When we got back, I had to wash the smelly garbage out.

And about two days later, Roy said, "Are you ready to go again?"

"Yeah," I said.

The plane leaped into the air again. I never flew the plane, but sat anchored by a seatbelt inside. Roy didn't do all the fancy maneuvering this time. He might have faded it or maybe executed a wingover, but nothing that made me sick.

When I attended college, I piloted a plane under the tutelage of one of our neighbors. We became great friends for life. Flying that aircraft fully-ignited the smoldering, flickering desire inside me into a burst of bright flames that reached high intensity when the United States Army sent me to military flight school.

First Flight Off a Grass Strip

At Texas A&M, in College Station, Texas, my next-door neighbor was an Air Force fighter pilot—Rabbit Hare. They named him Rabbit because of his last name. He was a short, stocky guy. Jean and I became close friends of Rabbit, his wife Virginia (we called her Jenny), and their little girl named Bunny. Their daughter had to wear glasses. If you'd take those glasses off Bunny, she'd grab them right back and put them on because she couldn't see without them.

Rabbit was raised in Amarillo. He would sweep out the hanger and do other chores. The man who owned the hanger said, "If you make a good hand, then I'll teach you how to fly." Rabbit got to be eleven or twelve years old and the old man taught him the basics of being a pilot. He let him fly around and take people places.

Years later, Rabbit became a fighter pilot. In Korea he had flown about a hundred missions.

He was stationed at Bryan Air Force Base, near College Station, when he was a 1st Lieutenant, and would fly training missions out of the base. When he was ready for Jenny to pick him up, he would buzz low over our street to let us see him in the cockpit. But as he approached the campus, he would roll the plane on its side and gain altitude to hide his plane's registration number. Then we'd go out and get him.

One friend of mine, another A&M student, had an Air Aeronica, a tail-dragger aircraft with limited instruments in it.

I said, "Hey, a pilot friend of mine wants me to fly that airplane."

Rabbit took me for a ride in the borrowed aircraft before he had me fly it. We went out east of my house to a big pasture surrounded by thick woods.

"I'm going to show you how to take off from a short strip," Rabbit said.

He gave it full throttle to go across the grassy field. The plane sounded like an old washing machine. But the trees kept getting closer. Rabbit kept its nose down, rolling faster, then he pulled back on the stick to rotate off the field. We almost scraped the trees with our landing gear. Rabbit went around and landed.

"Bob, that's not what I want you to do," he said. "Now you take off."

"Alright."

I held the brakes and put full power on it. The Aeronica made the washing machine sound. I released the brakes and we rolled over the grass strip. I did whatever Rabbit said.

"Keep her down," he said. "Keep her down." Then when we got enough speed, "Pull up. Over the trees."

A narrow gap opened between the trees and our plane.

We made it. That was my first flight training with Rabbit. He checked me out in the pasture, which was my flying experience in college.

About a year later, the Air Aeronica owner's wife was pregnant. He planned to fly her from A&M to the hospital in Hempstead to have the baby. His old Air Aeronica creaked into the sky and close to halfway on the trip, over Highway 6, the plane's prop came spinning off.

He glided, landed on Highway 6, and steered the plane into a roadside ditch. He helped his wife out and thumbed a ride to the hospital. No problem. His wife had the baby at the hospital. While nurses cared for his wife, my friend borrowed his parent's car and gathered us to go find his prop.

We cruised back and forth on Highway 6. The farmland was low and flat, with fields of cotton and other crops. You could still see the Aeronica's tracks where my friend had run off the road.

"Now we've got to back up about a hundred and fifty feet." He pointed. "The prop is out there in the field somewhere."

We spread out, five or six of us, and walked through the area. We found the prop sticking straight out of the ground like a knife plunged into the earth.

We pulled the prop out, came back to the owner, and he inserted it on the spindle. Using an old wrench, he tightened a few bolts and the airplane was fixed. His brother, who had come with him, drove the car home. After we positioned the plane on Highway 6, he fired up the engine, lumbered into the air, and flew back to College Station.

That was the airplane we flew over the trees on my first

flight. It didn't have enough power to do anything. Imagine if we'd lost the prop trying to climb over the trees. That would have been the end of the story.

CHAPTER 2

The Early Military Years - Germany

I entered active duty service in the U.S. Army in August 1954. To give you a little perspective, the Korean War concluded in 1953, though North Korea poked and prodded South Korea in a tense peace that still exists today. The next major war the U.S. fought was in 1959, in Vietnam. The war escalated from the presence of U.S. Special Forces advisors working in South Vietnam to full-fledged battle in the '60s and the first part of the '70s.

However, when I entered the U.S. military, we were in the midst of a cold war between the Soviet Union and United States. NATO forces in Europe became our allies in this great endeavor. The threat of nuclear war permeated our culture as the United States and our allies who stood for democracy faced the Russian Bear who promulgated communism as the world's best government. In the midst of these building tensions in Europe, I began my military career.

Our First Overseas Move to Germany

When I entered active duty service, the U.S. Army sent me, accompanied by my family, to initial training as an Artillery officer at the Artillery Basic Course, Fort Sill, Oklahoma. The school lasted seventeen weeks. From there I was sent to Nuremberg, Germany, unaccompanied until I could get concurrent assignment, which meant I had a place for Jean and our child to live.

In Nuremberg, Germany, we stayed in the 55th SS Barracks for about a month, where Hitler had housed his Shuttzstaffel (SS), the paramilitary organization for enforcing Nazi ideals and his personal protection. From Nuremberg, the only place we could shoot our eight-inch howitzers was at the Grafenwoehr training complex. We practiced everything, including emplacing artillery and firing the guns.

My unit moved to Schwäbisch Hall, Germany. After I had a set of quarters, which is military lingo for your apartment or house, I received concurrent travel approval for Jean and the family. She flew over to Germany after we had been separated about four months.

But Jean's trip to Germany was a traveler's nightmare.

Clarence, Jean's daddy, put her and Pam on a train in Fort Worth, Texas. Jean was young, about twenty-two. Pam, our first born, was eight months old. They rode the train to New York. At the train station soldiers greeted the women coming to be deployed overseas and drove them to an Army hotel in New York City. The soldiers would arrange transportation for them to the airport and for an overseas flight.

Everything was fine until a hurricane stormed into the Atlantic. Jean and Pam couldn't fly out.

Jean had to spend one week in New York. When the hurricane finally cleared out, the soldiers put her and Pam on a flight to Frankfurt, Germany's major international airport.

When Jean arrived, she had fifty cents in her pocket. She had spent all of the money in New York eating, feeding the baby, and taking care of the basics of life. When dependents, which are the soldier's family members, arrived in Frankfurt, U.S. Army soldiers welcomed them to Germany and accompanied them by bus to an Army hotel in the city. From there, the welcoming unit's role was to move the dependents to their soldier's duty station.

"Mrs. Gosney," a sergeant said, "your husband is stationed in Schwäbisch Hall, Germany."

"Yes, I know." She nodded.

"We're going to have you board a train to Schwäbisch Hall," he said.

"No, you're not either."

"What do you mean?"

"I'm not moving from this place," she said. "You get in touch with my husband and tell him to come here to pick me up."

Well, we were in the field in Grafenwoehr, a big training area.

One of our pilots flew an L-19 airplane, one of the two the Battalion had, into the training area and delivered a message to me. The pilot had been back in our Battalion Headquarters and the Executive Officer had told him, "Take this note to Lieutenant Gosney. Fast."

The pilot handed me the paper, which read, "Get your tail over to Frankfurt quick and get your wife. She's vexed."

I rushed to see the Battalion Commander.

"Go get her," he said.

The L-19 flew me back to Schwäbisch Hall. I grabbed my car, drove to the hotel in Frankfurt, and met Jean and Pamela.

At that point, Jean didn't have a dime. But the Army fed her and provided food for Pam—milk and other baby food.

"We'll go back to our house in Schwäbisch Hall," I said.

The roads were pretty clear, making our trip uneventful.

Our quarters were on the economy, which meant we stayed in an apartment in a German house. There wasn't enough housing on our military installation. After a few months, we moved to an apartment on post.

The German house on the economy was two-story, with an apartment downstairs and one upstairs. We lived in the upper apartment. Coal stoves heated the building. We had one coal stove in the apartment living room, the kitchen, and in our bathroom. The bathroom stove heated the room and the water for bathtub.

When we arrived at our home, I settled Jean and Pam into the apartment. The same day, without delay, I went back to Grafenwoehr. My battalion officers and sergeants took care of everything.

"Don't worry," the Executive Officer said. He had stayed behind. "I'll take care of your family."

The first night Jean stayed in the house she heard a bang, clang, clang, bang. Because of the cold, icebox-like weather, Jean had Pam in bed with her. And she smelled a horrible odor. Then, upstairs came these footsteps. And it was ten or eleven o'clock at night.

A man smoking a cigar eased open the apartment door, shuffled to the living room stove, and tossed in a little coal. He banged and clanged with his tools and shut the stove door. He left the living room for another room.

Jean was terrified.

The man was just an old German making his rounds, stoking the stoves to keep the heat blazing for the whole night.

At daylight, Jean got the biggest butcher knife we had, which the Army furnished along with other kitchenware, and slept with that knife.

The Battalion Exec visited later to check on her and Pam. Jean told the Exec what had happened.

"Don't worry about him," he said. "That's Hans. We have a contract with him to keep the coal stoves burning for you to keep the house warm, since it's wintertime. He also sets coal by the hot water heater in the bathroom. When you want to take a bath, all you have to do is put four coal briquettes inside the water heater. Shake it, wait about thirty minutes, and you'll have hot water."

The Exec's explanation satisfied Jean. When I returned from Grafenwoehr, she and Pamela were doing great because the battalion sergeants and the Exec even brought food to her. They'd go to the commissary, buy food, and give it to her because she didn't have any money yet. She was well-cared for.

When I came out of the field at Grafenwoehr, Lieutenant Colonel Brown, my Battalion Commander, said, "You fly back on the L-19 to be with your wife."

I flew back to be with Jean. She had mastered all the necessities of life in Germany. And once she knew the old guy banging around the house and smoking horrible cigars was the coal man, she could manage. We had nice living accommodations. The coal created steam heat for radiators, which warmed the rooms. The radiators had coverings on top to prevent someone touching them and burning their skin.

East of Schwäbisch Hall was a displaced persons camp. The American ladies decided to gather clothes to take to them.

A big wire fence surrounded the compound and the women bustled in the main gate carrying big boxes, which they had to leave. Then the displaced ladies hurried out and rifled though the boxes. Jean and the other wives felt disappointed. They had no control over the distribution of the clothes to make sure everyone got something.

But Jean met a lady in the camp from Poland. Her name was Tutu. She was the sweetest lady. Jean got her out of the camp because she gave her a job. And Tutu became a close companion for Jean.

Germans walk a lot. Like the local custom indicated, Jean often said, "I'm taking Pam for a walk."

Tutu, Jean, and Pam strolled around the caserne for fresh air and exercise. As military duties permitted, we would take off on trips lasting up to a couple of weeks, and Pam would stay with

Tutu and her husband. It didn't take long before Pam became fluent in German. That tickled me to death. She would talk to Jean and me in English and to Tutu in German. Tutu was a fine, wonderful lady.

Tutu and her husband invited us over to their apartment as friends. For social gatherings, they had cognac and sweet cakes. I didn't drink much of the cognac. Tutu idolized Pam, who soon learned to walk when she was nine months old.

My son, Mark, was born in a German hospital in Stuttgart, Germany, while I was stationed at Schwäbisch Hall. The hospital staff was all German doctors and nurses. Because I was away on military exercises, our terrific friends in the American housing compound watched over Jean for me during that time.

Lieutenant Colonel Hugh Brown—My First Battalion Commander

As a brand-new Second Lieutenant (2LT) when I arrived in Germany, I was assigned to the 291st Artillery Battalion. Lieutenant Colonel (LTC) Hugh Garden Brown was the commander of the 291st when I was a battery commander. Battery command was a Captain's position, but I filled those shoes as a 2LT. The reason for such a junior officer to assume command was because the Army had a reduction in force which caused a loss of many officers in those years. I used my sergeants (SGT), a few in officer positions, to accomplish the mission. They did a great a job for me.

LTC Brown was a short fellow, about as tall as my wife, Jean, who is five feet three inches. He sported a little mustache and had a sweet wife. But he was a "tough little banty rooster." He had been an Artillery officer in the 1st Infantry Division, known as the

Big Red One, in WWII. And for some reason, he and his wife fell in love with Jean, who as the wife of a young Lieutenant became the President of the Officers' Wives Club.

LTC Brown was a bugger. I liked the guy, but he was strict as could be. Our unit had towed 8-inch howitzers. In the dead of winter, LTC Brown insisted we clean the tires and make the guns look spotless. During the rainy season there was mud everywhere. When we got to our bivouac place, we'd wash the tires on the 8-inch guns. Even in freezing rain or snow, we still had to scrub those tires. I found a little secret I didn't tell Brown. If you put a little gasoline on the washrag, towed the gun on line, and then wiped off the tires, they'd shine like brand new. LTC Brown loved me since I had the outstanding battery in the Battalion.

He was a stickler for detail. We had an alert every month where we had to load out all our equipment, guns, and supplies. At that time we had nuclear rounds for our 8-inch guns. We already had our gear preloaded to a degree, but on alert we would pile everything on board, including our nuclear ammunition, and conduct a vehicle road march to our assigned war position. We practiced our mission to defend Western Germany at the Fulda Gap, which lay on the East German border. For alerts, we didn't go near the border at the Fulda Gap. Instead, we'd go through the road network in Germany for realistic training. But the tracked vehicles for towing the 8-inch guns would destroy the gravel roads. The Germans charged the U.S. government a pretty penny for restoration fees.

At our destination, we would find firing positions and establish the positions as though preparing for battle. A battery of four guns could stretch over a mile in what we called a battlefield maneuver. For nuclear firing safety purposes, the U.S. Army required that much separation from one gun to the next.

On one of those alerts our convoy wound its way back home. A soldier driving a truck with a water trailer hitched on clipped a tree, knocking a huge dent in the trailer's side. When we pulled into the motor pool, the first thing I saw was the dented water trailer.

Everybody called the sergeant driving the truck Red because he was redheaded, red-faced, and had red skin. Red was my motor sergeant.

"Sergeant Red," I said, "we've got to get the dent out of the water trailer."

"No problem," he said, "I'll get the dent out."

"Well, start doing it." I jerked a thumb at the trailer.

He crawled inside the tank—empty of water—and sat, pushing on the crumpled metal.

LTC Brown came around to look at the vehicles and equipment, carrying his chrome steel swagger stick. He was ticked off.

"Lieutenant," he said to me, "who is destroying my property?"

"We had a little trouble with a tree out there, sir."

"I want to see the driver of this thing." He whacked that water trailer hard.

Red leaped out of the trailer.

"Who was the S.O.B. that did that?" Red's hands rubbed his ears hard. "My ears are ringing."

"Sergeant," LTC Brown said, "I've got to tell you: I'm the S.O.B. that did that."

LTC Brown and I burst out laughing. Red kept massaging his ears, but he didn't smile until later. LTC Brown was a good Battalion Commander.

You have fun when you get the opportunity to work with 8-inch howitzers and Army vehicles. We'd deploy to Grafenwoehr to fire our guns for about two months at a time. We stayed at Graf in old Army barracks.

For one deployment we moved into four barracks buildings. The first one we opened, I kid you not, was infested with bed bugs. I'd never seen a bedbug until then, but you could see them everywhere. We hauled all those mattresses outside.

"Burn 'em," LTC Brown said.

We lugged the mattresses into a clearing, built a huge stack, and set them on fire. The Army post had an Engineer Detachment that came out and sprayed the buildings with insecticide to kill the pests. A French outfit had lived there without complaint for a couple of months before we arrived.

Instead of sleeping on beds, we pitched pup tents and set up camp while the engineers eradicated those daggone bedbugs from the buildings. LTC Brown was patient, but he gave me jobs to pass the time even though I was a battery commander.

"Bob," he said, "I want you to build a big concrete pad here and I want the shape of the pad to be like the Big Red One. Paint it to be an exact replica of this." He handed me a Big Red One patch. "Make it extend from here to there." He pointed out a pretty large area.

Well, I wasn't an engineer or anything. But I got a man to do the job.

"Dig all around this outline about six inches deep," I told the soldier.

He put rocks in the trench he had dug. Then he mixed concrete.

"You're going to make the pad look just like the Big Red One." I showed him the patch.

We did it. Thank goodness it wasn't wintertime. The concrete hardened in no time. We used concrete paint from our supplies and created an amazing Big Red One pad. The end result was an exact replica of the patch.

Afterward, I talked to the Post Engineer outfit's leader who had worked on delousing the barracks.

"We painted that Big Red One concrete pad," I said. "What can we do to keep the weather from ruining it?"

"Oh." The Detachment Commander tapped his jaw. "We have something to do the trick. It's clear and will go over the painting like plastic."

The plastic-like glaze worked great. Once it dried, we had a big ceremony with the Group Commander present. The Group Commander was a guy named Rex Rawie. He was a strapping full Colonel (COL) who carried a pearl-handle pistol like George Patton. LTC Brown served together with COL Rawie during World War II.

Colonel Rex Rawie and the Demotion

We headed home on another motor march from Grafenwoehr back to Schwäbisch Hall, looking forward to a good rest. Our unit's 8-inch guns and tracked vehicles, called tracks for short, moved on the German trains. The soldiers and wheeled vehicles melded together into a truck convoy.

COL Rawie, a dramatic guy, made an unexpected appearance during our convoy's return. A couple of trucks had minor mechanical failures and broke down. Red, the motor sergeant, drove the two-and-a-half ton maintenance truck. He got the trucks running and sped up to rejoin the rest of the unit.

COL Rawie's jeep pulled next to and paced the motor sergeant's truck as it returned to the convoy. Rex himself stepped out onto the running board of the moving deuce-and-a-half on the maintenance man's side.

"Maintenance Sergeant," he said with a bellowing voice, "you're a no good blankety-blank. These trucks are falling out of the convoy everywhere."

COL Rawie reached into the cab, tugged on Red's shirtsleeve, and with a big knife he slashed the stripes right off. He left a gaping hole where stripes had been. He was dramatic, I tell you.

"Colonel," Red said, "I can do a lot of things, but I can't make sparkplugs appear out of nowhere. We don't have sparkplugs."

With that, COL Rex Rawie stepped back off the moving truck onto his jeep. He raced away.

I spotted the action, drove next to the truck, and made the driver stop.

"What'd he do, Red?" I slid some fingers into the flapping hole. "Good grief, he cut your whole sleeve off."

"Yeah," Red said. "That sorry devil reduced me."

"He didn't reduce you," I said. "It was a bunch of show. You're still a Sergeant First Class. Don't worry about it. When you get to the unit, have the sewing lady for our Battery sew new stripes on your big jersey. But I don't think she can fix this." I spread the fabric wide.

Red did what I told him.

I don't think COL Rawie ever did anything about it. He was trying to be like General George Patton.

Educating Private Jones and Other Personnel Headaches

When I became a battery commander of 291st Field Artillery Battalion, I was a 2LT filling a Captain (CPT) position. LTC Brown, my Battalion Commander, had to relieve a CPT who commanded Bravo (B) Battery before I took over.

He called me to his office to break the news.

"I'm making you Commander of B Battery," he told me.

To take command of a unit, one of the first steps is to conduct a change of command inventory of all the unit's equipment. The outgoing commander had to ensure I had every piece of equipment listed on the books. The guy I relieved wasn't any help, even though he could be charged for each missing item. But we got it done.

I assumed command of the battery and had great sergeants. However, I had one sergeant who got promoted and demoted so many times I lost track.

"You'll get court-martialed if you don't straighten up," I said.

I was glad most of the sergeants were terrific. I remember

Ted King. He became a Master Sergeant in 1943, which meant he had twelve years at that rank. In those days, Master Sergeant (MSG) was as high as you went for enlisted ranks. The Army didn't create and promote to the next higher rank, Sergeant Major, until 1959. To have a guy of his caliber was incredible. And your MSG was the key to success. They organized and prepared the troops to deal with wherever we had to deploy, like the Fulda Gap.

MSG King prepared the sergeants and when it was time to go, he'd come see me.

"Sir, everything's ready," he'd say.

Having a non-commissioned officer (NCO) like MSG King made battery command easy. I never worried about the NCOs business because he kept them in the palm of his hand.

MSG King stopped by my office one day.

"Sir, we have a new replacement," he said. "I have to tell you one other thing. He can't read or write."

He brought the young soldier into my office. I'll call him Jones.

"Private Jones reporting, sir." The soldier saluted me.

"Jones, pull up a chair," I said. "I want to talk with you. Did you go to school?"

"No," he said. "We had a mule that died. I decided not to go to school to take care of the family."

"You can't read or write?" I examined his appearance. "How did you pass the tests to become an Artillery Soldier?"

"This nice recruiting Sergeant took 'em for me."

"Okay." I turned to MSG King. "Send me the Chief of the

Detail." The Chief was in charge of the firing battery.

"This is Private Jones," I said when the Chief arrived. "He is a sharp, sharp man. Your problem is teaching him to write his name. I want you to teach him to sign his name for his payroll signature."

I ran the payroll operations. I'd pay, the soldiers would sign their names, and I would count out the cash for them.

"Put your sign here," I would say the first four months. "I'll initial it."

"No sir," he would say, "I'm gonna sign my name."

He leaned on the desk, used his pencil, and scrawled his name at the pace of a tortoise, s-l-o-w. I took a close look, and sure enough, he could sign his name.

"Now you need to get somebody to help you read and write better," I said.

In time, he could write a letter home.

Now thinking was the hardest thing to do on the 8-inch howitzer. First, the soldier had to stake in the gun. Second, he had to line up the telescope sight. A person had to be able to read and think numbers. Man, that soldier could figure numbers in a flash. He wasn't stupid. He'd never had the opportunity to read or write.

I was proud of him.

But we had other personnel headaches. One guy in the battery stunk all the time. He smelled like a garbage pit. He wouldn't shower. To fix the problem, I had the men soak him and scrub him with lye soap. The First Sergeant (1SG) got him clean.

"You have a choice," the 1SG said. "You'd better take a shower every day or we'll do this again."

The troops never had to bathe him again.

Nuclear Rounds for U.S. Artillery In Germany

As I mentioned earlier, we had nuclear artillery rounds in the 291st Artillery Battalion. We stored them in the basement of an old battalion headquarters building. The German Luftwaffe used the structure when they had been stationed there.

Schwäbisch Hall was where the Luftwaffe flew the first German fighter jet. To keep the planes hidden, the Luftwaffe had constructed underground tunnels and hangers. Big tunnels connected all the buildings together.

The U.S. Army engineers welded the tunnel entrance doors shut and flooded them to keep young GI's from getting inside. The government's rationale for the action was that too many trip mines and bombs had been left in the tunnels and the water would keep curious American military personnel from getting through the doors.

Whether they were flooded or not, I'll never know. No water ever seeped out of the iron doors. By the runway, huge ramps also went underground to hangers, the storage area for the German jets. The pilots would bring them out on the big grass runway for takeoffs and landings. The underground facilities included aircraft maintenance bays.

Due to its previous purpose, our caserne had a large area to take our howitzers to practice moving them and setting up.

Our nuclear tipped eight-inch artillery rounds were for the Fulda Gap scenario. We received them in Pirmasens, Germany. The entire pickup operation was top secret, yet we'd drive a convoy there with two MP jeeps in the front, two in the rear and a deuce-and-a-half with a steel box on it. Three other two-and-a-half-ton trucks that looked alike drove in the convoy. We'd navigate through a German town and the MPs would blow their sirens. Following them came three two-and-a-half-ton trucks with more MPs blowing their sirens. I'm sure the Germans thought, "Good grief. I wonder what they're hauling in those trucks?"

The convoy wasn't secretive in the least. We'd retrieve the nuke rounds and store them in the basement. U.S. Army engineers put steel over the basement windows to meet stringent physical security requirements. Bright fluorescent lights lit the basement to provide the best illumination to assemble the warhead.

And I never thought much about what it would have been like if I was told to use those rounds. It never occurred to me.

If my commander had said, "Fire. Fire your nukes," I would have used them. It's something I didn't dwell on. A failsafe procedure ensured the order wasn't given unless it was necessary.

Two different senior officers had to issue the orders and the senior officers were not in the same unit. Each nuclear equipped unit had a codebook. Orders were given by a code, which told you to fire. The person giving the order information listed the names of the two senior officers. The person also confirmed the National Command Authority was the source of the approved order.

The Crazy S2 at Schwäbisch Hall

Our battalion had a lot of visibility because of the "secret" convoys that got our nuclear ammunition. At our base in Schwäbisch Hall we executed security duties with great attention to detail since we stored our nuclear ammo there. Around the caserne we had a big fence held in place by concrete poles. From the street, the Russians snapped pictures of the compound's interior. Part of the WWII peace treaty let them take pictures through the fence, but they had no access. The Russians couldn't come enter the caserne, but they did their best to understand what our operations were like.

We had a Lieutenant who filled the Intelligence Officer billet, the S2. He decided to test my Readiness Standard Operating Procedures (RSOP). The plan, when put into motion, simulates how the unit will shoot, without firing the rounds.

My battery had a big kid for a soldier with arms extending way below his knees. He could carry a round of ammo on his shoulder. That's incredible, because one round weighed 204 pounds.

While we prepped for a drill, a Russian sauntered into my RSOP area. The soldier saw him coming. He jumped right on top of the Russian and pounded away.

"I got that son of a gun," he said. I'm paraphrasing. The actual language was a might stronger.

The Russian wasn't a spy, but the S2 LT masquerading in Soviet gear. He was upset.

"Any fool who sneaks on this caserne dressed as a Russian had it coming," I said. I went to see my Battalion Commander, LTC Brown.

"Bring the young man in here," LTC Brown said. "I want to congratulate him."

About two months later, the S2 lifted off in an H-13 helicopter, crashed into a hill, and killed himself. What an unfortunate ending for an American officer.

A Squab Cell Transforms a Worthless Soldier

Our field artillery (FA) battalion was the first battalion in Germany to receive nuclear weapons for the 8-inch. The unit's original guns were 155-towed artillery, which we turned over to the German Army. The 8-inch towed howitzers were a big improvement. The gun could shoot a 204-pound projectile 18,510 yards and strike a tank size turret with every round.

Not only did the basement of our FA battalion have welded steel over the windows, the entry door was a heavy steel door with an oversized lock. The large room inside provided sufficient space to keep the nuclear rounds and the squab cells. We had about thirty of each.

A nuclear round by itself was harmless. To make it lethal, you had to insert the squab cell. The internal process for detonation is highly technical, but not necessary to know to complete the assembly process.

The squab cell screws into the round. The worker seats the cell where the fuse would rest. The soldier places a steel screw covering on top, resulting in the whole assembled device measuring three or four inches. In this configuration, the ammo is armed and dangerous. When the fired round impacts, you'd get a nuclear explosion.

I had one soldier who was absolutely worthless. I called him in one day.

"Would you like to learn how to assemble an 8-inch nuclear round?" I said.

"Yes, I would," he said.

"Okay." I signed some papers. "I'm going to put you in the training program."

Soldiers with this specialty skill wore white covers over their shoes and white smocks over their clothes. I had my sergeant dress the guy and stand him where I could look at him. "Not bad," I thought.

He became the best assemblyman we ever had. I believe he thought he was a surgeon because the assemblymen wore gloves and other clean-suit gear. He was amazing. He had been working on the gun section, hauling 204-pound salvo rounds, but now he was the best soldier to handle this complex technical task. He became the leader.

Artillery Gun Emplacement and the Outstanding Artillery Unit Award

Commanding and operating an FA Battery required quick thinking, accurate judgment, and detailed emplacement. The process was intense.

Alerts occurred once a month. On an alert, you went out—lock, stock, and barrel. We'd arrive at our destination near the Fulda Gap, maneuver our guns, and emplace them. Higher headquarters would spot a battery a mile for our four-gun front, which meant our Battalion's front of guns ranged five miles in width.

Because of the sensitivity of handling nuclear rounds, I spent hours racing from one end of the line to the other in my jeep. If a soldier armed a squab cell by neglect or accident, we'd have a nuclear burst.

We would be in the area about five or six days. At ENDEX, the end of the exercise, we fell back in echelon across the Rhine River, going over on barges. On the other side of the river, we would settle into positions again.

During alerts, the U.S. Army would have alerts for our dependents and civilians. Citizens had to throw everything they needed into cars and leave the kill zone in six hours. The Russian Bear was a world power and the U.S. wasn't taking chances. We were in the middle of the Cold War.

During my command time, President Eisenhower had reduced the size of the Army. That reduction of force removed most of the officers from our unit, leaving the unit commanders and the enlisted men. The FA Branch also instituted a test of towering proportions for the NCOs. If the NCO failed he was out of the service. Since most of my NCOs were veterans of the Korean War, after the test I lost 95 percent of the battery's NCOs.

My 1SG was the sharpest guy in the world, but he was gone because he couldn't pass the daggone test. President Eisenhower was the one who started the whole thing.

We needed leaders in our units to maximize our efficiency and function. An FA battery of over a hundred soldiers is split into sections. There are fifteen men in an eight-inch howitzer gun section. We had four gun sections, a communications section, a firing battery section, a headquarters section, and an ammo section, which hauled all the rounds. Those rounds weighed 204 pounds each. The ammo section brought all the projectiles and charges from the ammo supply point, offloaded them, and prepared them for use.

My unit had the M-1, a massive 8-inch gun towed by an M-8 tracked vehicle. The crew drove the gun into position, unhitched it, relocated the M-8, spread the M-1 gun wells and let them fall to the ground.

The crew's first task was to orient the gun in the proper direction, aligning it with the other guns in the unit. Unit leaders completed this with a tripod-mounted aiming circle stationed a short distance to the front or rear of the gun line and adjusted according to a surveyed point on the ground. Once the aiming circle was ready to "lay" the guns, crews rotated the sight of their gun towards the aiming circle and aligned the gun with the common direction of fire. Then the gun is "laid." The crew used aiming posts to adjust the gun during firing missions.

Aiming posts were metal tubes about 5-feet in length. They looked like javelins or thin barber poles and were set equidistant from the gun, normally to the front. A soldier gazed through the gun sight at another soldier who had run out from the gun with the near aiming post, waving him left or right until he had the correct spot. The soldier jammed it into the ground. They repeated the process with the far aiming post.

In my day, we had a Lieutenant in the Fire Direction Center tent. We didn't have the luxury of lasers and automation the FA units use today. Our LT used several slide rules for manually computing firing data for the guns: deflection, elevation and charge. These computations accounted for the wind, weather, rotation of the earth, and weight of the projectile. Another soldier used pins and a large protractor to mark the results on a large, flat firing chart. Once he had the firing data computed, the LT announced it to the guns over the wire-based landline that the commo section had rolled out.

While the deflection and elevation were being set, the crew loaded the ammo. First, they loaded the projectile followed by the required powder charge. We had different charges for the gun: green and white bags. Each equaled a certain amount of explosive power for specific distances. The LT called out the information and the gun crew raced around the gun section, loading the projectile and charge, and setting the deflection and elevation using the aiming stakes.

For elevation measurement, the soldiers used leveling bubbles and graduated scales on the gun. The soldiers leveled the bubble, fixed the azimuth on the aiming stakes, and prepared the required charge. Next, two soldiers lifted the projectile onto a loading tray behind the breech. Four other soldiers used a ramming rod—a large metal pole—to seat the projectile in the tube prior to adding the charge.

Soldiers called forward observers reported where the rounds landed and relayed adjustments for subsequent rounds. After you fired the howitzer, a soldier with a big, water-soaked swab thrust the end into the breech. The water removed any residue and ensured the gun was safe to load the next round.

We ran the gun drill like clockwork. I had a great crew, but I had corporals or three-stripe buck sergeants running the gun. I was the only officer in the battery. The Firing Officer (FO) and the Chief of Survey were buck sergeants. I had a sergeant in each key job. My forward observer was an exceptional buck sergeant. Another top-notch buck sergeant led my wire team, which laid switchboard wire.

Higher headquarters ran my battery through tests to make sure we met the standard. Seventh Army officiated over all the battery tests. The Eighteenth Artillery Group gave the test. Units with good scores moved to the Corps level. This included writing and executing field orders. We took a battery test at the Group level and my battery was the best in the Group.

At the next higher level, VII Corps, I came out on top again.

Then we went to the highest level, Seventh Army, commanded by Lieutenant General (LTG) Clarke. Seventh Army was the current Army headquarters in Europe. At that level, B Battery did the field test while evaluators from the Seventh Army headquarters observed as the umpires. We had to create a five-paragraph field order that told the sergeants and soldiers what to do, what was expected, and where we were going. My senior sergeant at the time was a Sergeant First Class (SFC). The evaluators graded you on every detail in the planning and execution phases, emphasizing speed, accuracy, and completion of mission.

We finished the five-paragraph field order delivery to the unit and sped to our positions. We took the guns and split them off, each pulled by big M-8 tracks. The unit would put in a gun position, then further down the front the surveyor would have surveyed in another gun, until I had all four guns completed. We used theodolite surveys to make it quick.

Once the gun was surveyed in and the aiming stake was at a survey point, we would plot on the chart in the fire direction center each gun's exact location. We could do everything in five minutes.

We would receive a fire mission from our higher headquarters. My battery could fire the fastest times of any unit in Europe, I was told. The fastest firing according to the manual was three rounds every two and a half minutes. We beat that time with room to spare. I had the battery crew fire three rounds on the Seventh

Army battery test in half the time with no staff sergeants, no senior sergeants, and no chief of firing battery.

They were good soldiers and we set the record.

When I got done with the Seventh Army battery test, my Battalion commander called me over.

"B Battery is the outstanding artillery battery in Seventh Army," he said.

Then the Seventh Army Commander, a three-star General, awarded the trophy to me, the Battery Commander, with seven of my sergeants present. The Corp Commander was LTG Uncles. I got a nice trophy. That and a dollar would buy you a steak. I gave the trophy to the sergeant who was my Deputy Commander to mount in our barracks because my sergeants did all the work.

ARTILLERY DUNCE

A funny thing happened when we received the outstanding battery award from the VII Corps level. LTG Uncles, the Corps Commander, came to our caserne to present the trophy. Uncles had a field jacket with sleeves that extended too far, covering most of his fingers. But he was a sharp General. When he arrived to present the award, I had the battery in formation in dress uniforms.

I had one soldier, an ammo man, who was dumb as a gourd. But he could sure hump ammunition. His long arms made it seem like his hands were below his knees.

"Soldier." General Uncles stepped toward him. "How do you like this battery?"

"Oh, it's okay, General."

"How about the food in this battery?"

"The food stinks, General. It's rotten."

I had one dud in the battery. General Uncles talked to one man—the dud.

The three-star had to know the soldier didn't look too swift. But he got a kick out of talking to him. I met General Uncles later for lunch.

"Sir," I said, "You picked out the most unintelligent man in our formation. He's an ammo pusher. He can lift and handle a two-hundred-and-four pound projectile himself. He's a good ammo man but he's not the smartest guy on the block."

"I could tell when I saw him," he said.

I realized he picked out the dunce in the battery on purpose, to quiz him and see how he'd do.

WWII Brodie Pilot Changes My Career Path

Later in my assignment to the 291st FA Battalion, LTC Brown departed for another position and the new Battalion Commander, Major (MAJ) Ejner J. Fulsang, Jr., took command. In due course, the Army promoted him to LTC. He was a Brodie hook pilot during WWII, taking off from and landing on troop carrier ships. A Brodie pilot was one of the most daring jobs in the Army Air Corps.

As a Lieutenant, Fulsang had flown the L-4 or L-5 Piper Cub.

These fliers were the eyes of the ship. The Navy didn't have Brodie pilots, which meant Army officers did the flying.

On the top deck of a cargo ship or tanker were two tall metal poles with gangplank extensions hanging over the side of the ship. Between the extensions ran thick wires parallel to the ship at least twenty feet out.

With the ship underway, sailors raised the plane by pulley on wire under one gangway and rolled the plane on the wire to the end of the extension. There, a sailor suspended in the air by a safety rope, transferred the plane to the plane launching wire with a bungee-like cord running through the Brodie hook centered on the top of the pilot's cabin. A sailor connected the cord to an arresting brake winch that pulled the plane back to the end of one extension for takeoff.

The Brodie hook's hinged opening faced forward. Although the pilot couldn't see the hook above his cockpit, he could see hook's stabilizing rod welded the top of the engine housing, right behind the propeller.

LT Fulsang would fire up the engine, getting it running hot, as the arresting brake pulled him back. At full throttle, the sailors would release the brake. The plane would buzz down the wire and the Brodie hook released near the second gangplank extension. Fulsang would fly off, complete his observation mission, and return.

He would aim the plane's Brodie hook at what looked like a practice baseball pitcher's net hanging from the ship's extension rods. Fulsang had to snag the plane's Brodie hook in the net to catch and lock onto a ring attached to the bungee-cord. The arresting brake slowed him down and he'd cut the engine. Then the sailors pulled him back on the ship.

The ship was moving during his takeoff and landing maneuvers, which made the task much easier. If he couldn't snag the Brodie hook to land and ran out of fuel, he'd have to ditch into the ocean. He flew as a Brodie hook pilot for the whole war.

LTC Fulsang became a pilot at sixteen, though the Army thought he was eighteen at the time. He became my Battalion Commander as a MAJ who had been selected for LTC while I was still B Battery Commander.

Fulsang would come to my office and visit with me. I'd be sitting across from him, talking about business, and all the sudden he'd throw me a golf ball or a tennis ball. I'd catch it, of course.

"What in the world are you doing that for?" I said.

"I'm testing your reactions," he said. "You have good reflexes. I want you to go to flight school."

The final day he talked about the decision, he came to my office again.

"You're going to flight school in San Marcos, Texas, to Gary AFB."

"Why?" I said, "I don't want to go to flight school."

"Yes, you do," he said. "Army Aviation is the coming thing. It's going to be big."

Gyroscope to Fort Sill

Our total time in Germany was two-and-a-half years with the

291st FA Battalion. When we left in February 1958, we executed a gyroscope move. The Army sent our whole unit personnel to Fort Sill. We would take over the weapons and all the equipment of another FA battery at Fort Sill and the Fort Sill guys came over to Germany and took over our equipment. That was a gyroscope.

CHAPTER 3
The Early Military Years – Moving Into Aviation

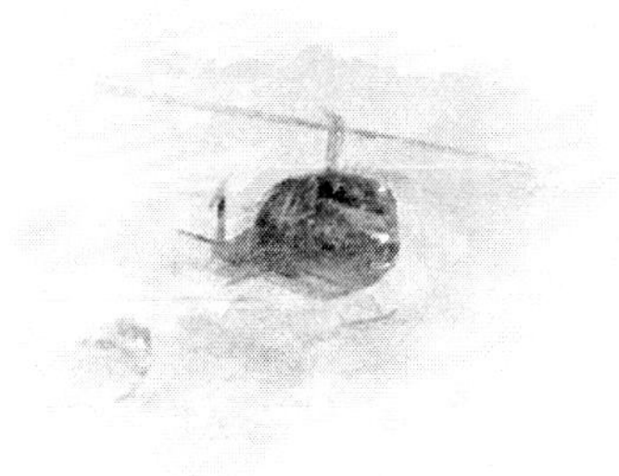

I didn't stay long at Fort Sill. The Army had a plan for my career, but due to LTC Fulsang's initiative about sending me to Army Aviation, he changed the Army's plan forever.

Soon after I arrived at Fort Sill in 1958, our daughter Molly was born. I also got orders to Fort Bliss to take the Air Defense Artillery (ADA) Officer Battery Course, which was about six months long. I left for school, moving the family with me.

The ADA course was a missile school at Fort Bliss, which taught us about the Nike Ajax Missile. It was awful. The instructors taught us how to build a missile, how they functioned, and how to fix them. That was the longest year of my life, even though the school was less than half a year.

I graduated the ADA course in 1959, which continued to be the year of schools for me. The Army approved my paperwork to become a pilot at Gary Air Force Base. Following my ADA course, I went to the Fixed Wing Aviator Course, a six-and-a-half month course, followed by the Fixed Wing Instrument Flight Course, which was about two months long.

However, I had a hilarious experience at the Nike Ajax School I can't forget...

Italians Fizzle at the Nike Ajax School in Fort Bliss

The Army limited attendance at the Fort Bliss missile school to regular Army Lieutenants. Since I was a Lieutenant, I received orders and reported to El Paso accompanied by Jean, Pamela, Mark, and Molly. We lived right next to the base in a tri-plex on the side of mountain.

I'll tell you flat-out—we never fired a single missile. Instead, we learned all the technology and circuitry, including how to compute the burst radius and other effects for a nuclear weapon. And those effects varied depending on the different KT (kilotons) the explosion would deliver, the weather conditions, and location of the burst.

When we had a test, the instructor would give directions like this:

"Trace this circuitry on the missile guidance board," he would say. Or he gave us other technical details we had to commit to memory.

I was bored to tears.

We saw one Nike Ajax missile fired. One. And it was a fiasco. The instructors put us on a truck and took us out to the range. The ADA battalion had an outfit out there to fire their first Nike Ajax. A plane flew overhead towing a target for the missile to shoot down. The teaching cadre sat us in bleachers looking at the preparations. When we looked close, I saw the battery firing the missile was an Italian unit. Their personnel went into a bunker-like building.

We watched, intent on every step in the process as they fired the Ajax missile. The missile fired, but didn't go anywhere for a few seconds. It vibrated, shook apart, and then we heard a swooshing sound as a majority of the rocket squirreled into the air. Straight up and straight down, right back onto the launching pad, exploding in a giant fireball.

We grinned, slapping each other on the back and laughing. I saw the Italian troops stick their heads outside the bunker, check for safety, and take a few halting steps. I guess they learned—you have to release the missile for it to leave the launch pad.

The Italians became the subject of several missile jokes. You can imagine. They had been there almost five months. Their whole job was learning to fire missiles, not memorize detailed circuitry and electronics. But I guess they got their money's worth.

Got Fuel? Surviving Fixed Wing Flight Training at Gary Air Force Base

I anticipated a better experience at Gary AFB, San Marcos, Texas, than I had at Fort Bliss. We had Air Force instructors and training. We flew L-19s, built by Cessna, which were powerful, single-engine aircraft with the latest instruments.

I enjoyed my adventures at flight school. For one mission I flew a cross-country flight to the Texas coast. My flight plan was to navigate to Corpus Christi, but difficult weather developed and visibility dropped to zero. I gained altitude and flew by my instruments.

I worked to stay on a heading to Corpus. In the end, the heading I used didn't take me to my intended destination. Still twenty-five miles away, I knew I needed fuel and landed at a small airport.

"Yeah, we got fuel here," the airfield manager said when I asked.

They filled the airplane's tanks and I paid cash out of my own pocket. My own pocket. I paid them twenty dollars.

Darkness fell and I thought, "I'll fly back from here to San Marcos."

I had my map and flew straight as an arrow to Gary AFB. I didn't get lost even though I was in rough weather, surrounded by clouds. I knew where I was when I could see. Not everyone could do that.

In my flight training class we had forty-one students. Matter of fact, we lost half of those, graduating twenty-one. The instructors issued one warning if you didn't perform to standard. One pink slip indicated trouble. Two pink slips and they flushed you out. A couple of Air Force instructors were bad guys. They thought forcing students to perform under greater and greater pressure would make them better pilots. But not all instructors thought that way and I got along with my instructors with no issues.

Qualification or Washout? Captain Waters, Kubis, and Ray Brown

Ray Brown was full-blooded Mexican and his father's name was Brown. His mother and father spoke broken English.

"Ray," I said, "how did you get the last name Brown?"

He laughed and gave me a fake story about his father's ancestry. But Ray spoke fluent Spanish and English. He told me, "I guess I've got gringo blood in my ancestry."

Ray was a dear friend of ours, a good guy. We lived side by side in duplexes in San Marcos. There were three of us: Ray, Mike Kubis, and me. Mike Kubis, another good friend, came from a Czech background. We drove to the air base every morning and came home together every night.

Each morning, Ray would come out and get ready to get in the car. The first thing he did was throw-up because of nerves.

"Ray," I told him, "get out in the street and do that. You're killing the grass."

I had to chuckle. Every morning he lost his cookies.

This happened during the preflight instruction, where you had to solo. The average time for a student to solo was four to seven flight hours. I soloed in four hours. Ray soloed in five hours. Kubis at twenty-five hours still hadn't soloed. Our instructor, an Air Force Captain, told Kubis he had two more flight hours to solo or he would wash him out.

In flight training you couldn't forget a single detail. One time Ray went through his landing procedures: fuel on the fullest tank, flaps at thirty degrees, and the other pre-landing checks. Ray did all with perfect precision... except he ran out of fuel on the short final because he hadn't moved the switch to the fullest tank. He forgot.

The instructor pilot took over and made a dead-stick landing. It had to be quick. The plane wasn't more than twenty feet in the air when the engine quit and the pilot's reflexes jumped into action to make the landing.

And, oh, the pilot, a Captain, was fuming mad. He went after Ray like a boar ready to gore his prey.

"When you do the preflight check," he said, "you look up and switch tanks to the fullest tank."

Ray had kept it on the same tank the whole time and it ran out...

I almost tore a muscle to keep from laughing out loud.

The Captain's name was Waters. Ray and I were sitting by the little portable tower. Captain Waters sat in the aircraft with Kubis, trying to get him to solo. And they came in, landed, taxied on the taxiway, turned the plane around, and got out.

"Kubis?" Captain Waters said. "Do you want to solo now or are you going to crash? Take off and go solo, for God's sake."

Kubis hadn't finished exiting the airplane. Kubis reversed what he was doing, settled into the pilot's seat, took off on the taxiway, flew a ninety degree turn across the runway, and went out of sight. He was supposed to turn the plane in a downwind baseline and come back. But, heck, he disappeared.

Waters stood there. "He's the dumbest Czech I ever..."

Kubis was nowhere to be seen and we were wondering what

was happening.

"See if you can raise that dumb, crazy Czech," Captain Waters said to the control tower. "He's gone. I don't know his location."

The tower controllers had binoculars. "I've got him in sight, sir," the control tower operator responded. "He's about five miles off the downwind."

"Is he heading this way?" he says.

"No, sir. He's still going out. We'll call him on the radio."

Kubis couldn't answer the radio.

Seconds ticked away, then we saw him. He turned downwind and he was way, way, out there. And he kept going.

"He's a good three miles off the runway," the tower announced. "Oh, he's turning across. He's turning base. He's turning final. He's losing altitude." They paused. "He's got his flaps down."

Kubis landed in a plowed, harvested cotton field. He drove the plane the remaining five hundred yards to the runway, then taxied to the tower where Waters stood.

"Kubis," Waters said, "why did you land in the cotton patch?"

"Well, sir," Kubis said, "everything was perfect. My rate of descent was perfect, my direction was perfect and I thought, 'Man, I might as well land where I am because it's so good.'"

That almost killed me.

"Kubis," Captain Waters said, "we're going go over there, fill this dang airplane with gas, and you're going take off again. This time, you're going take off on the runway, not the taxiway. You'll take off, you'll gain altitude, turn right...and I mean turn right on the downwind...then you're going to turn right on base, right on final, and you're going to land on this runway. Do you understand?"

"Yes, sir."

Well, he did. He landed and made a perfect landing.

Waters threw up his hands in disgust.

"Kubis," he said, "I hope you'll live long enough to enjoy flying."

The light came on in Kubis. He became one of the best pilots we had. When we flew tactical maneuvers, you'd have thought Kubis had already been in tactical flight school. The light bulb turned on twenty-seven hours later for Kubis. Waters would have washed him out, but Kubis was such an honest guy.

"I know, I know," he'd say. "I'm gonna get it, I'm gonna get it."

And he did.

I ran into him in Vietnam. He flew choppers there as a Loach pilot—the guy who descends to search for the bad guys.

"The light bulb came on that day, didn't it?" I asked him.

"Well," he says, "it really came on when he told me I was to land on the runway."

I chuckled. The runway was the one missing piece of instruction.

He told me: "Everything was perfect. The flaps were perfect, the rate of descent was perfect..." Yeah...except he was five hundred yards out in a cotton patch. Hilarious.

But you know, thank God he had Captain Waters, because he

was a nice guy—strict as all get-out, but a good heart.

Kubis and the Intentional Martini

Kubis is the one who started me drinking martinis with Ray Brown.

I had never drunk a martini in my life. Every day when we'd get home from flight school in the evening, Kubis would call us since we lived right next to each other. Kubis made the best martini, but I didn't know that at first. I had never tasted a martini.

He handed me one and I sipped. It tasted great.

Ray Brown and Kubis would throw their martinis down the hatch.

"Kubis? I said. "Do you do this every night?"

"Well, yeah."

"That's why the flip you can't fly an airplane." I laughed. "You're drunk."

"Oh, no. I sober up."

"Kubis," I said, "I recommend you not drink this stuff. Maybe one."

"Oh, it won't hurt me."

"Yeah," I said. "But you have to pilot these aircraft, you know. You're hanging by a thread."

After Ray and I had soloed, Kubis wasn't qualified yet.

Ray and I flew around rectangles on the ground for coordination, an essential skill to fly a single engine prop. Glancing at your instruments, the pilot has to keep the ball in the center while doing the rectangles. Both feet were working to keep the ball where it belonged.

Kubis flew rectangles, too. There's not an instructor or another student with you most of the time. Kubis could do it alone, but getting down wasn't easy. Taking off was no big deal, but his landings were a problem.

In the end, after soloing, his light bulb flashed on in his head and he could land like a pro. I was tickled. Kubis was a nice guy. Married to a beautiful girl. And he even had a little accent.

"By golly, by golly," was a typical Czech-inspired phrase.

Kubis was one of many reasons why I enjoyed flight school.

Learning the Tactical and Instrument Phases of Flight Training

After the Fixed Wing flight training, my next assignment was at Fort Rucker, Alabama, where I learned tactical flying. We were in the air day and night, landing in tight, confined areas. The instructors taught us how to stress those airplanes close to and beyond their specifications. Most aircraft couldn't stay aloft under those conditions. My remaining classmates and I didn't have any problems with the dangers.

We still flew the L-19 Cessna. In tactical flight training, I learned how to land in what they called the double S.O. I had enough room to bounce the plane in and stop it before crashing into the trees. Taking off from the same cramped space the first time made my blood pump a bit faster. We did tactical landings on roads. In those maneuvers, the pilot had to wheel land the aircraft, go around the curves in the road, and put the tail down, along with other tricky moves.

After the tactical phase, we went into flight training with instruments in the weather. The course seemed like several months. I didn't have any problem there.

I flew with my buddy, Ray Brown, and an instructor. Each student had to follow an accurate track on the Omni radar, the Automatic Direction Finder (ADF), and other instruments.

We also flew all the way to Atlanta. We used the normal air routes or airways, filed flight plans, and completed all the required paperwork and actions to make instrument approaches in to Atlanta, which is a pretty busy airport. I made about eight or nine instrument approaches in to Atlanta in those fixed wing aircraft.

Instrument Flight School and Pappy Jingles

In the fixed wing instrument flight school, the first plane we used was a big Beech aircraft. It had a single, huge engine. The engine was a powerful thing and the windshield extended pretty high. Then the Army decided to use the De Havilland Beaver aircraft, another single engine plane, but bigger than the Beech with better instrumentation.

The instruments for the aircraft were pretty crude. The plane had a round antenna on the top for landing—for coming in on a specific heading. The antenna was a low frequency contraption that made a sound like di-dit, did-ah, di-dit, did-ah. The di-dit was the letter A and the did-ah was the letter B. An A meant turn left and a B meant turn right. You had to listen with care. The only reason you knew you went over the broadcasting station for a landing in low-visibility weather would be an aural null. You might hear di-dit, dit-ah, di-dit, did-ah, di-dit, did-ah, and then for a split second, you'd hear nothing—an aural null.

At that point, you reduced the power knowing the plane was on track for a straight approach. You descended, broke through the cloud cover, and landed. However, with the massive roar pulsing out from the engine, you had to listen as hard as possible through the headset to hear the signals.

At times, Ray Brown would be in the back, intense concentration wrinkling his forehead as he listened over engine noise and looked out the window. He'd jerk my seam strap to let me know I was over the Omni station. When I'd be in the back seat, Ray would be in the front like I had been, trying hard as heck to hear the aural null. The minute I saw the station I'd pull his seam strap. When either of us was the pilot, we couldn't see what was happening because the pilot had to wear blinders to keep him from seeing the station. The pilot had to hear the aural null. A couple of times I got it, but that was the fickle finger of fate. Separating those tones from the noise in the cockpit was almost impossible.

We had different types of approaches to airports: ADF (Automatic Direction Finder), also known as NDB (Non-Directional Beacon), and GCA (Ground Control Approach). For the ADF procedure, I'd tune in an ADF station and crank my antenna to get a heading to the station. Once I had it, I'd follow the heading into the airport. As I flew over the outer marker of the airport for the ADF, I'd hear a dit, dit, dit...which meant I was over the outer marker. Then I'd start my approach to land.

A few days we flew in cloudy weather conditions out of Fort Rucker, Alabama. When daytime conditions were low visibility, flying became mighty rough because you'd have to bust through towering cumulus clouds. When you were in the clouds with no visibility, you were in the soup.

One instructor I had was a guy name Jingles, Pappy Jingles. If you asked him, he'd been a pilot for a hundred years. And he was a rough talking guy. Your task could be tracking an Omni station to approach on a direction of 191 degrees. If you flew off track one-half a degree or drifted in the wind one-half a degree off the station, he'd raise all kinds of heck with you.

Once I was shooting an instrument approach into Pensacola, Florida, on actual instruments, using the ADF. I had hit the outer marker, still in the soup, but I was taking the heading on the instruments. Pappy Jingles started cussing me out because I was off about a half a degree. I corrected and then he said the bad word. It's an expression I won't tolerate.

"You dumb S.O.B.," he said. "Can't you keep this plane straight?"

It hit me like a lightning bolt. I went in and landed.

The minute the tires hit the tarmac, he said, "Take off."

I shot up. A touch and go. We were back in the soup. I picked up my headings, did all the right things, and got on the airway. He hollered at me all the way back.

Ray Brown was sitting in the back seat.

"Oh, God," Ray said to himself, "I'm glad I'm not in front. He's giving Bob hell."

And Pappy Jingles was.

We landed at Fort Rucker, Alabama. I taxied the Beaver to the parking spot, pulled in, and shut it off. I got out, went around the plane, opened Jingles' door, and I dragged him out.

"You S.O.B.," I said. "If you call me that one more time, I'm gonna beat the living daylights out of you. You understand? I'll whip your tail all over this runway."

"I didn't mean it to be personal," he said.

"What did I tell you?" I said. "I'll whip your tail all over this place if you call me an S.O.B. one more time."

And man, he became my personal friend. I was ticked. Calling people names isn't good training when you're in the middle of a cloud, shooting approaches, climbing out of the soup, and getting headings.

I was steaming mad. I kept it in until I got the aircraft stopped. And I beat him around a little, I guess. I opened the door and jerked him out of his seat. He was near sixty years of age. I was probably twenty-three or twenty-four. But I was mad.

I flew with him for the rest of my instrument course and he never, ever called me an S.O.B. again. He never hollered or cussed anymore. Ray Brown was mighty appreciative because before that Jingles would get in the instructor seat and lambast him.

Learning to Fly Several Aircraft at Fort Hood

After all the flight schools, I went straight to Fort Hood, the Great Place, and flew fixed wing aircraft for a year.

I was stationed with the 502nd Aviation Company, part of the 2nd Armored Division. We did the unit's field aviation. After setup of the camp, we flew wherever the mission required. There was not even a runway, but a grass field we'd use for takeoffs and landings. We'd have our guys pitch tents out in the field to support the Division. We flew troops and personnel, including the General. After I became a helicopter pilot, I flew the General a lot in the Huey. I was also an instructor pilot in the Huey.

Jean and I had three children at the Great Place: Pam, Mark, and Molly. We had a wonderful family life and made many friends.

I seized the opportunity to fly several aircraft, but the standouts are the Grumman Mohawk, the De Havilland Otter, the DC-3, the Huey, and the De Havilland Beaver with pontoons on it for water landings. The single-engine Beaver provided a new qualification for me: qualification for landing on water. We used some of the local lakes to train.

Student Pilot Ejects Instructor from Grumman Mohawk

The Grumman Mohawk had ejection seats in it. Grumman built it for close-air-to-ground support, including the weapon systems. The company later upgraded the equipment and hung Side Looking Aerial Radar (SLAR) on the plane. To make matters worse, Grumman attached several other devices to the airframe. When the plane first came out, it would leap in the air at takeoff, but after attaching the extra equipment, the aircraft lost its show-off performance, crawling into the sky at a miserable rate.

I flew the plane out of Fort Hood after going to the factory, where I received my check out ride and qualification. We flew the aircraft from the factory to Fort Hood where we kept six of them.

Before the company upgraded the equipment, I piloted the aircraft for close to a year. The Mohawk is a sweet flying airplane, acrobatic in every sense.

A friend of mine came to Fort Hood and got checked out in the aircraft. He took off at the little airfield on post, lost an engine, and did everything wrong. He didn't compensate for the dead engine nor did he feather the engine prop. The Mohawk rolled onto its back. When the plane was almost upside-down, he ejected, which shot him into the ground, killing him. You can never think of a flight as routine.

Then Fort Rucker had several Mohawks. The unit was checking out its pilots. The student pilot sits on the left side and the instructor pilot (IP) sits in the right seat. The ejection handle for the pilot was on the left side of the seat and for the passenger the handle is also on the seat's left side. To activate the system, you yank the handle back and boom, you're ejected from the plane.

Well, one student pulled the wrong handle and shot the instructor through the canopy.

Of course, the IP's chute deployed. The setup included a chute that pops open if you're unconscious. The parachute landed in a pasture and the student pilot flew the Mohawk back and landed. He had to explain his missing instructor.

"I ejected the IP because he gave me emergency procedures," he said.

Instructors always tell students when they're doing an exercise.

"It's simulated," the instructor would say. "This is a simulated emergency procedure."

"You're right engine is going to die," the IP had said. "It's out of fuel."

When the right engine died, the student thought it was an emergency. Instead of pulling the handle to eject the canopy, he pulled the instructor's handle and shot him out through the top of the canopy. It's a wonder the IP didn't get killed by the propellers on the plane. They're close to the cockpit.

That was one ticked-off instructor when the crew went out, retrieved him, and brought him back in to Fort Rucker.

You might wonder if the kid made it through. The answer is, "Heck, no." The incident finished his career. The unit put him in front of a flight evaluation board and took his wings away from him. Which was the exact message to send in that circumstance. Faulty reactions as a student could result in death in combat as a pilot.

Hauling Tactical Nukes at Fort Hood

I was on the "Hey You" list as a duty officer to escort nuclear rounds to their destination. There were several officers who could pull this duty, but one day I was lucky and headquarters selected my number.

We flew in the back of a modified C-124 Globemaster II. The aircraft carried special radar in its nose. The old four-engine airframe never arrived at the destination with four engines running. The aircraft would regularly lose one engine somewhere. The types of malfunction were endless.

I was designated to appear at the nuclear storage site on Fort Hood and escort the rounds to Biggs AFB, El Paso. Fort Hood had a large number of ammunition bunkers, a couple containing tactical nuclear rounds. Giant tunnels connected the bunkers and today parts of these areas are used for offices. Biggs AFB had a Strategic Bomber Wing. Our nuclear site also delivered tactical nukes overseas to various places.

As the Lieutenant escort, I'd carry classified items in a briefcase strapped to my wrist. Inside were the orders, destinations, and quantities being transported. When you dropped off the munitions, the receiving unit had to sign lots of paperwork.

It's my educated guess that the nukes we carried were tactical because I took the weapons to U.S. Air Force Bases. I never knew the exact details about the nuclear warhead, but I had all the information in the briefcase. Upon arrival at our destination, nuclear surety officials would meet us to establish control the minute we landed.

I'd take up to ten rounds and go to five different locations. We also transported squab cells, the part inserted into a warhead to make it nuclear capable, which were in a separate container from the rounds.

Mastering Helo Piloting and Avoiding Pogo-Sticking

I went from Fort Hood on temporary duty to Camp Wolters, Mineral Wells, Texas, for Helicopter School, which was about nine months.

But before I went to flight school, I spent time in a helicopter at Fort Sill, right after we gyroscoped back from Germany in 1958. I rode with a guy who had been a helicopter pilot for four or five years. I wanted to get a feel for piloting a chopper to ensure I had the touch.

In helicopter school, I flew a lot of different helicopters at Fort Wolters and did well. I had the knack of flying a chopper.

"You got it," the instructor said when I completed my first few hours of training.

As a chopper pilot, your feet and hands are in constant motion.

The throttle is on the collective, which sits to the left of the pilot's seat. The collective looks like a hand brake with a thicker handle that rotates left and right. The rotating part is the throttle. On the end of the handle are several switches. The collective controls engine speed, altitude, and the angle of attack of the main rotor blade. The collective makes you go up, down, or hover.

The cyclic pitch control, another stick extending from the floor of the chopper to a place in between your legs, controls the tilt (pitch) of the main rotor. The cyclic has a handle on the top portion, like a grip on a handgun. If you push the cyclic forward, you go forward, to the rear (aft), then you go aft, and so on.

The pedals at your feet control the rear rotor on the tail, which keeps you pointed straight, left or right on a flat plane. The flight term is yaw.

It's a lot to control all at once, but a good chopper pilot can do it all from muscle memory without a thought.

Here's how I approach piloting a chopper. Roll the hand

throttle on the collective to bring it to sixty-four hundred RPM and hold it. Then you pull up on the collective and the chopper rises. And if you want to go higher you'd pull the collective up further. To go forward, you push the cyclic forward. If you're hovering and you want to turn left, you press the left pedal and the chopper rotates to the left or vice versa to go to the right. When you want to land, you pull the cyclic back a little to hold the nose high and lower the collective.

To hover, once you're at the altitude you want, you'd bring down the collective to keep it in the neutral position, and you can hover there for hours.

Putting all the actions together for a smooth flight takes a bit of experience.

At Camp Wolters, the helicopter school instructor would go out for training with you. At about a thousand or eight hundred feet, he'd cut your engine off, and you'd have to make an autorotation. In an autorotation, you have to lower the collective to descend which keeps air flowing through the main rotor. As you near the ground, you try to land in a normal fashion.

We started with an H-23, called the Bubble. The helicopter was light—a single-seater. You could carry a passenger in it, but the instructor didn't fly with you after you qualified. We were at a place called The Devil's Tail.

The area was about the size of a two-car garage. You had to bring the chopper in and make an autorotation to land in the little clearing. Then the instructor let you start the engine, come to a full hover, and take off. Next, the instructor soloed you. I soloed after two hours of flying the Bubble.

One time in the Bubble, I landed out at a little patch of land near Camp Wolters. A cloud of giant grasshoppers swarmed, jumping around the pierced steel plank (PSP) area, which was a rapid-deployment parking area.

"I'll just smash some of those bugs," I thought.

I descended and began smashing the insects with my chopper's skids. The next thing I know, I'm pogo-sticking, which is bounding up and down while losing control. It's a progression that worsens until you wreck.

"Dang, what am I doing?" I thought.

I raised the Bubble and got it flying again.

When taking off, a helicopter's rotors don't fly until you get translational lift. You can feel the vibration as you start forward movement and feel the rotors produce lift in cleaner air. Typical airspeed for this is about 16-24 knots. Once you're in translational lift, then you're flying and not merely hovering over the ground. I worked to achieve translational lift and gained altitude.

"Man," I said to myself, "that's the dumbest thing I ever did, killing those grasshoppers with those skids."

If you concentrate on the grasshoppers, you're pogo-sticking in less than a minute. I can laugh about it now, but it was a close call.

Our instructors advanced us to a more aerodynamic chopper. The helicopter had a pointed nose, a lot more power, and several other equipment upgrades.

I graduated as a qualified helicopter pilot at Camp Wolters and flew back to Fort Hood. At the Great Place, instructor pilots checked me out in the H-19, a pretty big helicopter, though it doesn't haul many people. When you land that chopper, you've got to be on your toes, because it will pogo-stick in a heartbeat. The

H-19 has rubber tires on it, with a tail wheel and two front wheels.

After the H-19, I moved into the H-34, which is bigger, hauls more people, and has much greater power.

Missile Tracking Into Mexico

As a Major, I was detailed with two other pilots to Biggs AFB. Our mission was to track the Honest John Missile being tested at White Sands, New Mexico. We flew the F-51 Mustang, a modified and modern version of the older P-51 Mustang of WWII fame.

Werner Von Braun was in charge of the missile project at Biggs. We met while I was at the base. He was the brain of the operation. The Honest John Missile they tested was twenty-seven-feet long and mounted on a truck for launching. It's a pretty big missile.

We'd get a mission briefing every day. There would be one airplane flying at a time. I was one of three pilots and we had two F-51 airplanes. I crewed one aircraft all the time and the other two pilots switched off to man the second plane.

At the briefing, the mission control staff stated the scheduled missile firing time in military time, like 0700. They provided impact coordinates on the White Sands range and a litany of other information. For testing, the warhead carried white cement dust.

I would takeoff in the F-51 wearing my pressure suit and climb to 25,000 feet. At altitude I saw the launch pad. When the crew fired the missile, I followed it to the impact area. The Honest John would go, and go, then strike, leaving a deathly white plume. I could see the whole thing.

"It hit," I'd call on my radio when the missile impacted.

Big markers covered the range.

"Marker 32," the technician would radio. "It hit within a hundred feet of marker 32."

"Great," I'd say. Then I'd go back and land at the airfield.

After one of the briefings, I reached 25,000 feet and saw the missile fire. The missile flashed into the sky, did a 180-degree turn, and headed right toward Mexico. I did a 180-degree maneuver to keep tracking from my over-watch position.

The Honest John sliced the air at an altitude of 20,000 feet, flew over Juarez, Mexico, and slammed into the side of the mountains at a treeless spot. The powder and the dust spurted up into the air.

Now, you're forbidden to cross the border into Mexico; therefore, I held back. I stayed high where I could verify. The technicians kept calling me, but I wouldn't answer. Then Werner von Braun got on the radio. I could tell by his accent.

"Sir," I said, "I'll be on the ground in five minutes and give you a full update."

"Oh, no." Ground control followed that with a few choice words. They got the message—something went wrong.

I landed, climbed out of the cockpit, and Werner von Braun's sedan rolled up with a few of the staff. I walked over to the car.

"Vell, Major," he said in his German accent, "give me the bad news."

"The bad news—the missile flew into Mexico," I said. "It impacted on the Sierra Madre mountain behind Juarez. There wasn't a tip or anything on it. But it smacked into the south side of the mountain."

The group huddled. Either they got permission to go into Mexico or decided to leave the wreckage because it would take a long time to find it. I don't know what happened.

But I was there for six weeks chasing Honest John missiles. It was a fun job.

I loved piloting the F-51. It has a throttle stop for takeoffs. You move the throttle forward and hit the stop. If the throttle didn't stop there, the pilot wouldn't have enough pedal control to overcome the torque and keep the plane from spinning around. You had to takeoff with the throttle at the stop. The minute your wheels are airborne, you can pop the throttle over and move it forward to gain more speed. Then you can open her up for a great flight.

Hurricane Carla Search and Rescue Mission

During Hurricane Carla, I took a whole company to Ellington Airfield in Houston in September 1961. We had five H-34s, five H-19s, and we had two H-37s, which were the largest choppers. I didn't pilot the H-37. It was twin-engine with gasoline reciprocating engines. Those choppers burned about a barrel of oil an hour. You always had to keep putting oil in the system.

We flew to Ellington during the hurricane, which tracked along Interstate-35. Visibility was low due to heavy rain and wind, but the whole company made it without problems. We started getting search and rescue (SAR) missions right and left. We found out while flying that the maps we used were almost worthless. Carla caused the coastline to move about a mile inland. What was supposed to be a cove in one spot was no longer there. It was gone.

And little towns were washed away.

In downtown Port Lavaca an ocean steamer was laying on its side.

One washed-out town had a single concrete pad left intact. An old man stood on the pad. I landed the H-34, shut it down, and got out.

"Sir," I said, "where can I take you?"

"No place." He looked me over.

"Where were you during the hurricane?" I said.

"Right here." He waved his arm at his feet. "Underneath this slab there's a basement."

"You're kidding. Didn't it fill with water?"

"Yes." He gave a slight nod. "It filled with water and I had about that much air." He held his hands a foot apart. "You're not taking me anywhere. I'm staying right here. This is my home."

"You haven't got anything," I said.

"Don't worry." His eyes locked on mine. "I'll take care of myself."

"Alright." I climbed back on the H-34.

He was an old salty guy.

We continued the mission. I sent choppers to as many locations as we could handle. We hauled food, medicines, supplies, and other articles to where the need was greatest. We scoured the area looking for victims. I'd send L-19 Birddog planes out to follow fences. We found a couple of bodies tangled in barbed wire fences. We had to recover the bodies and take them back to graves registration at Ellington.

CHAPTER 4

NEXT STOP—VIETNAM I

Before I reached Vietnam in 1966, I spent the years after Hurricane Carla filling assignments the Army directed me to take.

First, I took the Armor Officer's Career Course, which was intended to prepare officers for command and operations in garrison and tactical environments. We reviewed all types of military actions, including how we would fight the Cubans during the Cuban Missile Crisis. There's never a dull moment in the Army.

Next on the Army's 'To Do' list for me was to go to Korea for a year as the Operations Officer for the 55th Aviation Company. In months, I became the Commander, 8th Army Aviation Company at Yang Do Po Korean Air Base, near Seoul, Korea. I flew General Officers around Korea, to include supporting the Eighth U.S. Army Panmunjom mission for negotiating with North Korea. I piloted visiting U.S. Congressmen and their wives to Tokyo, Japan, for shopping and sometimes to Okinawa for a tour. One of the dignitaries I flew was Bobby Kennedy and his wife Ethel.

My follow-on assignment was to the Command and General Staff College at Fort Leavenworth, Kansas. Our instructors taught us the information we needed to function as staff officers at the Field Grade level, which is Major through Colonel.

The final assignment before my first tour in Korea was as the Executive Officer, 3rd Howitzer Battalion, 6th Artillery Group, Fourth U.S. Army at Fort Sill, Oklahoma. I had a bird's eye view of the numerous challenges facing a U.S. Army field grade Commander. I soaked up and integrated the details into my thinking as the world situation heated up. The Vietnam War grew in the eyes of our national leadership and the U.S. Army responsibilities increased enough to cause my selection as part of the team in the war. Operational combat, while scary to some, was the mission I had prepared myself to execute with complete success when the time arrived.

My First Impressions of Pleiku, Vietnam, as a Combat Chopper Pilot

My first tour in Vietnam was from 1966 to 1967. The first four months I served as the operations officer in the 119th Aviation Company (Air Mobile) in Pleiku, located in the area near Chu Pong. If you traveled north from there, you would be at Dak To, the central Vietnam highlands. When the Commander,

119th Aviation Company rotated back to the States, I became the Commander of the unit for eight months.

Pleiku, our base of operations, had the normal seasons—rainy cold that chilled your bones or hot, muggy air filled with dust and earthy smells. I didn't mind sweating on hot days because we could cool off a little in our chopper. And the optempo (operational tempo) kept us busy. We were in combat daily, averaging ten piloting hours a day. We did "hot refueling." For example, we'd fly into Plei Me or Du Cho. The ground crew had fuel bladders ready to pump. You landed, kept the chopper engines running, and let them fill your tanks. That was a hot refueling.

When we flew operations in the mountainous areas, the maximum fuel our helicopters could take was eight hundred pounds of fuel. This amount allowed you to navigate the mountains, haul the full load of ten troops, and take any action necessary. If you had to extract somebody, your chopper would be light enough to haul more people. If the chopper carried a full load of fuel, you couldn't carry more than five troops. Now, eight hundred pounds of fuel didn't last long when you're soaring high in the mountains. I staggered the number of choppers operating and refueled them often. We'd sometimes have the ground crews move the bladders a little closer.

Every day in Vietnam we started off at five o'clock in the morning. Because I had an air-mobile company, I would brief the mission to my ground forces platoon leaders the night before the next day's execution. We'd leave at five o'clock sharp and start the assault at first light. For all of these missions, I piloted the command and control ship.

Hitting a PSP with No Tail Rotor at 60 Knots

One time I flew a chopper back without its tail rotor.

Enemy bullets slammed into the rotor and blew it off. The impact was at the base of the tail rotor. Since the whole mechanism was gone, I struggled to counter the rotational forces. My senses ratcheted up a notch.

I knew the tail rotor was missing because I had to use a heck of a lot of rudder. I was going above 60 knots (69 MPH) when the bullets hit. My crew chief looked out, hanging out the door.

"Sir, the tail rotor's gone," he said.

"Okay," I said. I kept calm though I felt my blood pulsing. "We'll maintain and land at 60 knots. Everyone hang on when I land." Standard procedures called for the crew to have their belts hooked on and to brace themselves.

For a while, it was touch and go. Landing at 60 knots is pretty doggone fast and dangerous when you're landing on a short PSP (Pierced Steel Planking) runway. But I needed the high landing speed to avoid spinning in circles. At a slower speed without a tail rotor, the helicopter would rotate.

PSP is a metal strip with holes in its construction for rapid assembly, linking strips together under the pressure of a combat situation. Our PSP runway had to be long because we had a terrible landing. The minute I smacked onto the metal at 60 knots, I bottomed the pitch, pushed the collective full down, and shut off the engine.

The chopper started a slight rotation at 45 knots. The greasy metal allowed us to glide at a slow spin. If I had landed on dirt, the ground would have caught the skids and flipped us. We spun around eight to ten times while sliding the length of the runway.

I got out, breathed deep, and looked around. The crew chief and gunner took longer—they were a little drunk from the spins. When the crew chief jumped out, he checked the damage.

"A few hours and I'll get you back in the sky," he said.

We had to replace two skids on the chopper because we wore them down almost to the support beams on the PSP.

The holes on a PSP are about three or four inches in diameter, three or four times bigger than a quarter. At the ends, the PSP hooks together like a railroad track. The engineers would place them, hook the planks together, and create a runway. They could lay a three-thousand-foot runway in a day. South of Pleiku, the Douglas A1E Skyraiders, air attack planes, operated off a PSP runway because setup is fast.

PSP runways are terrific. When we had the rainy season, the sloppy weather didn't make any difference to a PSP runway at all. Unlike their dirt counterparts that became unusable because they had turned to mud, the PSP was a dependable landing site.

The Plantation Ultimatum

Higher headquarters in Vietnam maintained pre-established free-fire zones. The zone would cover a large area. Anything moving in the zone was fair game as a target. It didn't matter who moved, including women, because they could be fighters as well.

When flying out of Pleiku, the flight path went over a plantation to the south, the Catecka Tea Plantation. You couldn't trust the plantation owners to keep out the Vietcong, because they protected their properties from war disruptions.

I flew into a village near the plantation one time. I wanted to talk to the village Chief, a short brown-skinned man, because I needed information about the presence of any Vietcong on the nearby Catecka Tea Plantation. An interpreter spoke for me.

"If you see anybody around the tea plantation," the interpreter said to the Chief, "let us know." He made sure they knew we were talking about the Vietcong.

"Oh," the Chief said. "Yes, okay."

I wasn't sure whether the Chief and his people would tell us or not.

A day later I got my answer.

I had several of my choppers fly over the plantation enroute to a mission. When they were within a hundred yards of the main house, the chatter of semi-automatic weapons filled the air and bullets smashed into a few choppers. They sustained enough damage to require a return to base for repairs.

When I heard that, I was ticked off. I piloted a gunship out to the plantation and planted it on the front yard of the French villa, the owner's house. My gunner hopped in the front to keep the engines running. I stormed toward the door, pounding on it as hard as I could. An older man, the French owner, swung the door open. He spoke English.

"Listen," I said, "I'm going to tell you this once. If I get another round fired at my choppers coming across this area, I'm going to blow this doggone house flat. Now you get it straight. I will blow your house to ashes."

"But this is not my fault," he said.

"Baloney, they work for you," I said. "They're out there on the tea plantation, hoeing and growing the plants to make you money."

He still didn't want to take the blame.

"I'll blow your house away. If you don't believe me, I'll get back in there and I'll fire two rockets." I pointed at my chopper.

"No, no." His eyes widened with fear. "I'll take care of it. I'll handle it."

"You'd better," I said, "because I'm not going to have my helicopters shot down when they're crossing this plantation."

That solved the problem. We never got shot at again. The Frenchman knew I would've gone back and demolished his house.

But I have to admire those Frenchmen. They had been in Vietnam since the French-Indo China war. They were beholding to the Vietcong and to the North Vietnamese for letting them do their business there. The French had to work with the bad guys to survive.

But I didn't care about that. I wasn't going to have my helicopters shot at by AK-47's when they flew over the plantation. We won that fight.

The Chopper I Flew in Vietnam

Now, the Bell Huey helicopter was the best helicopter I've ever flown in combat. That airframe can take a heck of a lot of beating and still fly. We had the single engine Huey in Vietnam, the UH-1 D model, built for medical evacuation and utility work.

The Huey was what we would call a slick troop carrier. On both sides you could mount 2.75 mm rocket pods with twenty-four rockets in each pod. The gunner or pilot could dump all forty-eight rockets at once into the enemy or you could fire two rockets at a time, two from each pod. We could vary the attack.

The rockets were high explosive rockets. Now a Dust Off chopper, which was an unarmed helicopter and only carried medical evacuees, was on a similar airframe, but they flew on the M model. The one I flew on was the D model.

Dealing with Montagnards: An Arrow in My Chopper and C-Rats

The Montagnards were a nomadic people. Their life consisted of slashing and burning to raise one crop of rice, then, with the soil depleted of the right nutrients to grow more rice, they would move to another place. The tribes-people didn't know whether they were in Cambodia, Laos, or Vietnam. They didn't care where they were. The main goal was to feed their families.

I had returned from a mission and my crew chief was doing the post-flight inspection.

"Hey, sir." He called me over. "Come here and look at our chopper."

I slid underneath, thinking, "Holy cow, I didn't know we got hit."

"Look there." He pointed with a tool.

I saw an arrow stuck in the belly of my chopper. I smiled. It was a Montagnard arrow. They used handmade crossbows. We had flown over a village and they must have fired at us. And I don't know why they would fire. They didn't care what side we were on, but maybe they got scared. They might have thought we were a bad-looking bird or something.

The Montagnards aren't educated at all. I know this because we were sitting in a little firebase where I had brought three other choppers along with me. I was talking to the commander. We were going to take soldiers and insert them in another location to relieve pressure on a specific unit.

While I was talking to the commander, a whole line of Montagnards walking through the firebase caught my attention. The women came first. They always walk in front and the men behind because if they get ambushed the women get killed and the men get plenty of time to scatter. Their clothing looked like rags and most were bare from the waist up.

They kept walking in through the unit and I was astonished to see the most beautiful little Montagnard young lady I've ever seen. She was gorgeous enough to be a TV model.

By now, I was sitting back in my chopper eating a can of C-rations. I motioned for the Montagnards to come over to me. I planned to give them C-rations because I had three cases in the choppers. They got the idea and trooped over to see me.

The little girl came over. She had nothing on but a covering hanging from her waist. It looked like a sarong, but her clothing was the traditional way the natives dressed. I thought, if you could take that young girl, scrub her with GI soap, and clean her up, her appearance would be striking. Her beauty would stop men in their tracks.

I don't have any idea how old she was. She looked to be fifteen or sixteen. But she was must have been younger; because when Montagnard women are into their middle teens they already have black teeth and other signs of neglect or aging. Her beauty even caught the attention of my co-pilot, another Texan.

"My Lord, look at her," he said.

I wondered how she could be with this group of people. Most of the women were about thirty, but looked like they were in their sixties. The Montagnard women age at a rapid rate because their life is tough.

I called the man I thought was the head of this group and I showed him we had C-rations for food. I had one of my sharpshooters, which is what we called the P-38 can-opener, and opened the can to show him how it worked. The leader didn't understand the demonstration. I finished opening the can and handed it to him, along with a plastic spoon. C-rations all come with plastic spoons. I motioned for him to eat. He took a couple of bites and his lips parted in a wide grin.

The girl beauty, whose intense gaze followed each move of my can-opening instruction, stood right next to him, but he paid her no attention.

"Let her taste it." I pointed.

With reluctance, he let the little girl have it. She took about two spoonfuls. Her eyes widened and glowed. She was happy.

They passed the C-ration around. It might have been butter beans or another vegetable.

I issued out cans to each individual with a sharpshooter. When I gave them to the young girl, I watched her reaction.

She looked at the P-38 and at the can, confused.

I knew she was starving hungry. I took the sharpshooter, punched a hole in the can, and began opening it with a few twisting motions of my wrist. I returned it to her.

She opened it the rest of the way herself, flashed me a big grin to let me know she had mastered the task. She ate every bit of the food in the can.

The rest of the Montagnards watched the action. I handed out more C-rations to them. I had about three cases to give away: one case in my chopper and two on the other choppers.

They came in the camp with nothing and left with three cases of C-rations. That amounted to about four cans per person.

The Montagnards carry small bags made of vines and other materials. They stuffed C-rations and sharpshooters in the bags and threw them over their shoulders. I felt confident the little girl would be able to eat the food, because she had caught on fast as to how you open a can.

I'm sure they all understood how to do it.

The group left by forming a line, the women in front and the men behind, and walked out of the firebase in a single file.

Chopper Down—Ammo for Ranger Unit at Plei Me

I got shot down at a Special Forces (SF) camp at Plei Me.

At two o'clock in the morning I was making my way back to Pleiku and a Special Forces guy heard my chopper. He called me on an emergency radio channel. I was coming back from a reconnaissance mission after working with another Special

Forces unit, planning an assault.

"Helicopter flying over Plei Me," the SF guy said.

"Are you calling me?" I said.

"Chopper Six" the radio crackled, "have you got any spare 30-caliber or 5462?"

Chopper Six meant he was calling the helicopter commander. The number '5462' was the last four of a national stock number for standard 5.56 ammo for an M-16.

"I don't know," I said. "Let me see."

"Have we got any spare ammo?" I asked my crew chief. "If we do, how much do we have?"

"We've got plenty," the crew chief said.

I keyed my microphone. "Yeah, we've got it. Do you need it?"

"I sure do," he said. "You can't land in here though, 'cause we're under heavy attack."

"I noticed a lot of gunfire," I said.

"Yeah. I think it's a North Vietnamese regiment," he said, "and we're running low on ammunition. Fly across and turn right, then you'll be clear. But come across low."

"Well, tell you what I'll do," I said. "I'll come over fast and low, and we'll kick it out the door." It was a dark night, but the moon provided dim illumination.

"That would be good," he said.

I skimmed across as low as possible, almost scraping antennas.

The gunner and crew chief threw out the ammunition. Then the helicopter wobbled and the engine sputtered.

"Holy Toledo," I thought as my mind flipped into high gear. "I've got to get this chopper on the ground."

The engine started to quit. The turbine made one heck of a racket. And I'm about thirty feet in the air. I could feel the RPMs bleeding off.

"Jiminy Cricket." Like a racehorse in a steeplechase, my thoughts leaped over the obstacles one at a time.

I tried to keep the RPMs high. I did a slow pedal turn, put the nose down, and hit the runway hard—a dead stick landing. I gripped the controls tight to get out of the middle of the camp, because the chopper blades could kill people. The runway landing was a good autorotation because the rotor slowed on the turn. I flared, pulled pitch and crunched down hard, but not too hard. I got her on the ground and everyone was okay.

"Get out," I said. "Get out of this daggone thing. Bring your M-60's and all the ammo you can get."

We exited quick. The gunners brought four tins of M-60 ammo.

My co-pilot looked like a Mexican bandito. He had a sawed off double-barrel shotgun and shotgun shells hanging over his shoulders like a bandolier.

At the end of the runway by the camp, we spotted a big ditch and scrambled in.

"You get on that end," I told the co-pilot. Then I ordered the others to their positions. "You two machine gunners get in the middle and spread out. I'm taking this end."

All I had was a .38-caliber pistol—the smallest gun on the battlefield.

We got into the ditch even though the SF camp was right next to us. We couldn't get in. The concertina-wire fences and mines guarding that fence encircled the camp for full protection, showing no entry point. Not to mention, we were in the middle of a battle. The SF soldiers fought like crazy, firing over us. They fired flares out from the camp, caught the North Vietnamese Army (NVA) on the runway, and mowed them down. My guys helped with their M-60's.

After scrambling into the ditch, we snuck away from the chopper because we didn't want the chopper blocking the view of the SF soldiers. We crawled to stay out of the line of fire and to get a clear shot at the enemy.

"Be on your toes," I said to my M-60 guys, "and if they come over here, you erase them off the map. I've got my .38 pistol and six rounds of ammo left. No extra bullets." I didn't realize how small my revolver was until I was under fire.

"I've got plenty of shots," my co-pilot said. "Buckshot."

We yelled to communicate and did all right. The battle raged until daylight. Though I was dog-tired, adrenalin-surges kept me alert. I couldn't get the attention of anybody in the camp.

Every now and then we had a lull. You could have heard a cricket chirping. Lying in the mud, I heard something. It sounded like someone crawling closer toward me. I couldn't tell for sure in the dimness, but I thought someone might be moving. A flare lit up the area and I stared, eyeballs straining, but I saw nothing.

"Somebody's coming." That nagging thought didn't go away.

For a while, the SF didn't fire any flares. I listened to myself breathe while on my belly, waiting, and holding my .38 pointed at the black space in the gaping ditch ahead.

Out of nowhere, a face looked at me. I squeezed the trigger and hit him right between the eyes.

He was an NVA soldier coming to kill us.

I stopped him cold.

He wore a backpack containing two old socks, with rice balls in them, and he had plenty of 7.62 caliber ammunition for an AK-47, which he had dropped.

Immediately I stripped the pack off him. I felt a lot better with the AK-47 in my hands. Better firepower and a good supply of ammo. Next to the AK-47, the .38 pistol was insignificant. I used the AK-47 the rest of the night and kept the rifle. In the mid-'60s, you could bring confiscated weapons home.

Most of the night, the NVA attacked and the base repelled them, shots whizzing over our heads.

We couldn't do anything until daylight for the simple reason that if you moved out into the open, the SF camp soldiers might think you're the wrong guy.

When daylight came, I attracted the Special Forces attention by yelling into the camp. The Captain in charge informed me he had Montagnard soldiers to make sure I didn't think enemy had penetrated the perimeter.

"Keep crawling through the ditch," the Captain said. "It'll come right around and I'll open the gate. Then you can hunker down, jump, and run in the gate. We'll lay a base of fire to cover you."

At last we got inside the camp. For the next three days our sole job was to fight like infantrymen. The NVA attacked all day followed by night attacks.

The North Vietnamese called off the attack for several reasons. The Air Force had sent A1E attack aircraft to drop napalm on them, spreading it over the area. A Lieutenant Colonel brought his Republic of Vietnam (RVN) Ranger battalion in from the north to counter-attack. Between the Ranger battalion and the Air Force pounding the living daylights out of the NVA with bombs and napalm drops, they broke and cleared. The Ranger battalion was key to breaking the siege.

But the A1E Skyraiders did a yeoman's job. The pilot puts the plane in a vertical dive, then puts dive brakes out. The plane dives vertically and drops the bombs straight down. Then the pilot pulls the aircraft out of the descent, makes a one-hundred-and-eighty degree turn, then comes in to drop the bombs again. The A1E carries the same bomb load as a B-17 carries, about eight thousand pounds.

The Skyraider is a single-engine airplane developed for the Navy, but in time the Marine Corps and Air Force flew the airframe. The plane met the exact needs for Vietnam combat operations.

As the NVA retreated to their base of operations, they tried to drag their dead, but couldn't. The A1E attacks were relentless the whole time. The NVA left about two hundred dead on the battlefield.

Another incredible statistic: at the Special Forces camp they didn't have any casualties while defeating a North Vietnamese regiment. None.

One reason for this exceptional stat is that the Special

Forces layout at Plei Me was a triangle, a typical SF defensive design. Bunkers covered the triangle's center. Soldiers fought out of perimeter fighting positions that covered every avenue of approach with their fire patterns.

The North Vietnamese didn't have artillery, but they had mortars. As long as the soldier stayed inside his defensive fighting position, he was pretty well protected unless a bullet made it through the shooting hole.

I would have loved to have Puff the Magic Dragon at the battle, but we didn't have any in our AO (Area of Operations). Puff was the nickname for the Douglas AC-47 D Spooky gunship, which had two mini-guns that fired from the side of the aircraft, decimating the enemy.

"Have you tried to contact Puff?" I asked the Special Forces Captain.

"Yes," he said. "But there aren't any in this AO."

A couple of days later, our unit retrieved us, but until we left we worked like infantry grunts.

I had a Chinook come in and haul out my chopper to our home base at Pleiku for repairs. The mechanic found the problem. A bullet had penetrated one of my engines, severing the fuel lines in the system. The mechanic had my chopper fixed and running in no time. We were back in the air in three days. All he had to do was replace the fuel lines in the engine. That was a freak shot. And the amazing thing is no other bullets hit the engine.

I still have the pistol, the rucksack, and the AK-47. If you put an AK-47 on full automatic, it will empty out the banana clip pretty quick. The weapon can shoot single shots and as a semi-automatic as well. The gun has a high degree of accuracy.

Working Hard in the Jungles of Vietnam

I hate to say this, but I enjoyed Vietnam, not for the war, but because I served with the finest soldiers in my entire military career. The draftees needed someone to lead them. I led them with firmness and compassion.

I worked harder in Vietnam than in any other military assignment. We worked nonstop, sometimes for incredible hours. Sweat flowed, choppers thundered, and we'd scour the hills for the enemy. I flew a continuous seventeen hours on one mission, never shutting down my helicopter as we dropped fuel bladders around the AO. We took quick breaks while ground crews hot-fueled our choppers. Several choppers were riddled with bullet-holes, but limped back to base.

I spent a great deal of my time on the ground with infantry troops. Under the jungle's triple canopy, an M-16 sounds like cannon fire. Apes squealed from treetops. The NVA had entered the war, but didn't admit they were fighting. Like the Plei Me episode, during my units' skirmishes on the ground, we killed NVA, around seventeen uniformed soldiers.

Mortars and Muddy Windshields

An infantry battalion was in an LZ (Landing Zone) and I worked a re-supply mission with them. As soon as I landed, the NVA on the high ground fired mortars.

The mortars exploded in the muddy clearing and the mud, filled with shrapnel, sprayed into the air covering my windshield.

I took off, seeing only one or two trees of the jungle's forest. I couldn't get translational lift to rise over the trees. I backed up and landed again. When I hit the ground, the crew chief hopped out to clear off the mud. I started for the trees again and still didn't have enough lift to get over them. I hovered even further to the rear, hoping for enough lift because the mortars were falling all over the place. The third time was a charm. We got translational lift and high-tailed it out of there. We visited this location five to six times-a-day. A few pilots got scared, but they did their job.

After the unit secured the area, one-hundred-and-fifty dead NVA lay in a pile. The decomposing bodies' pungent smell reeked and we formulated a disposal plan.

I had a Chinook haul in four barrels of JP-4 jet fuel. We used thermite grenades to set them off. The two hundred gallons ignited and burned the bodies. We repeated the process, but the unit had to abandon the firebase. We couldn't burn enough of the remains to relieve the stench.

The Two Brothers Story

I had two brothers in my unit named Fuller. One of them, Mike, asked me to send a chopper to bring his brother, Dan, a 1st Cavalry Division soldier, to visit before an attack. When Dan arrived, his clothes were rotting off him. My soldier, Mike, dressed his brother in some of his clothes. During the night the enemy killed Dan in a mortar attack.

After the attack, I received a telegraph saying the enemy killed my soldier. However, when I went forward, I found my troop and realized it was Dan, not Mike, who had died. I had orders cut to send the remains home, but Mike, my soldier, wanted to see his brother's remains.

I accompanied him to the morgue in Saigon and made Mike stand outside. There were five hundred bodies in the morgue, whole or in pieces. I told the Mortuary Affairs (MA) officer I wanted to see Dan Fuller, but he had been misidentified as Mike.

"How can you say the body is Dan Fuller?" the MA officer said.

"His brother, Mike, is my soldier, and I just left him standing outside."

"Then why is he wearing Mike Fuller's clothes?" the officer asked.

"His own clothes were a mess. Mike gave him some of his own."

The officer nodded and led me to the body. Dan Fuller's wounds were significant. He was un-recognizable from the mortar attack. He was clothed with his brother's uniform and the name was still visible.

Casualty Affairs for the Army had already notified Fuller's parents and wife of my troops' death. I had them send another telegram to the family to correct their first notification. They had to know Dan Fuller had been killed, not my soldier.

"You don't want to go in there," I told Mike.

"Okay, sir." Mike said. "He had a wife and a brand new baby."

"Go home and take care of them," I said. I put him on a plane and he left the Theater of Operations for home.

Two years later, Mike Fuller visited me at home in the United States. He had married Dan's wife, adopted his daughter, and became a police officer in Dayton Ohio, all of which indicated his outstanding character.

Company Rescue—3rd Brigade, 1st Cavalry Division—Hal Moore's Brigade

Hal Moore, the Lieutenant Colonel who led the 1st Cavalry (1st CAV) Lost Battalion as they fought in the Ia Drang Valley near the Chu Pong Mountain had become the 3rd Brigade (BDE) Commander. After his military career, Moore wrote the book, *We Were Soldiers Once...and Young,* about the Lost Battalion experience. Little did I know we would meet in Vietnam's jungles.

I was cruising south on the Vietnam coast. We had been operating with the Korean Tiger Division north of Phu Bai. Our operational area was far up the coast. We returned late at night, flying low in a V-formation during the rainy season. The rain was a thick, heavy, downpour leaving us poor visibility. I had seventeen of my choppers with me. That's how many choppers it took to do our job.

I was always changing my radio frequency to the command frequency of the units I piloted over or came through. I checked my little book and dialed to their command frequency. While flipping through the channels, I heard a young officer who was in dire need.

The officer and his unit were isolated, out of range of our U.S. artillery, and out of mortar rounds. He was under heavy attack and was calling his BDE for help.

The BDE Commander, who was Hal Moore, had told him the 1st CAV, Air Mobile Division said they couldn't fly to assist. COL Moore told the officer he would do his best to get a unit there to relieve him. But Hal doubted he could make that happen, because the single avenue to get a relief there was by air. Happy Valley was a VC (Viet Cong) controlled valley.

I heard the conversation and called my choppers.

"Okay," I said. "Let's land right here."

I had them touch down in line. Then I called the officer who needed support fast.

"Where are you?" I said.

"I'm in Happy Valley," he said, "and we're in bad shape here. We're under attack and all I've got left is M16 ammo. I have no more mortar rounds. I need help."

"Give me a clue as to where you are."

"I'm twenty miles north of Brigade Headquarters."

I knew the location of the BDE HQ.

"You are twenty miles north of there?" I asked to confirm.

"Yes."

"I'm close to you," I said. "Tell you what you need to do. When you hear me coming down, start flashing a red light at me."

Charlie Barryhill, a Chief Warrant Officer Four (CW4), and I were instrument qualified. The rest of the nineteen-year-old pilots were not. I gave a brief rehearsal on how to navigate by instruments in a Huey.

"Remember your attitude indicator is going to show you are descending when you are climbing," I said. "Watch your attitude indicator, your vertical climb, and vertical descent meters."

I finished telling them what indicators to monitor.

"We're going to line up in trail," I said.

I called Qun Nhon. Their air controllers had GCI (Ground Control Intercept) radar to vector Air Force airplanes. I called the radar station and told them my location.

"When we rise to two hundred feet," I said, "we'll be in the weather. I need you to track me, help me make a left turn, take me over the mountain, and let me down in the valley. I know you'll lose me when I'm close to the valley because it's mountainous terrain. "Can you do that?"

"Sure," he said.

I turned on my Identification Friend or Foe (IFF) transponder, which beeped a signal he could see on their radar.

"We've got you," he said.

I had my choppers lift to a hover.

"Turn your navigation lights on," I said. "If it's not too dark in the clouds, you'll be able to see each other. But if it's dark, maintain your relationship to the guy in front of you and in the back do the same thing. We're going in there and extract this outfit."

I told the flight if anyone did not want to go they should RTB (Return to Base) and I wouldn't fault them. One by one they called off their call signs.

Every one of those pilots said yes. I was proud of them. I put them into trail formation and had them turn on their running lights.

"If any of you guys have trouble," I warned them, "make a slow descent and go home."

We lifted and were in the soup at two hundred feet. White fog. We continued to climb as the controller turned us toward the mountains, took us over them, and helped us descend. We broke out of the clouds near two hundred feet in the valley. The co-pilot searched for the ground and the pilots behind him started calling, "One's out, two's visual..." until all seventeen popped out. The ceiling varied between two hundred feet to as low as fifty feet. Believe it or not, all the nineteen- and twenty-year-olds did a good job. They didn't crash or have any problems.

I saw red lights blinking a quarter-of-a-mile from where I was.

"Can you hear me?" I said. "Can you hear the choppers?"

"Yes, sir," came the response. "I can hear them."

"Where are your men?"

"I've got them here next to the rice dikes. We're still under attack."

"Do you have anyone in the ridge line?"

"They're all bad guys."

"This is what we'll do." I paused a second. "We'll come in and land on the rice dyke. You mount up and we'll fire our machine guns on the right and left sides. Have your men keep their heads down when they approach us and we'll keep the VC off of you."

Hovering under the weather, we started taking fire.

"Shoot back," I told the flight.

The gunners fired out each side of the aircraft. I called for a cease-fire as we came too close to friendlies.

That was a long fifteen minutes because they had three KIA (Killed in Action) they had to collect. But we kept the enemy at bay with our machine guns. With fourteen slicks, troop-carrying choppers, I had fourteen machine guns firing one side and fourteen firing the other side.

The bad guys fired at us, too, while we loaded.

I sat my helicopter by the Company Commander who guided me in with a spotlight. Young soldiers ran out of the bush. The First Sergeant had organized the soldiers into groups of eight for each chopper. The Captain got in my chopper. His 1SG loaded the soldiers in the other choppers and put the KIAs in the last chopper. After I thought they were all loaded, I glanced at the CPT.

"Have we got everybody?" I said. "Your 1SG is in the tail. Here's a radio. Talk to him and verify no one's missing."

"Everybody's aboard," the 1SG said to the CPT.

"We're good," the CPT told me.

I told the gunners to open fire with the machine guns. I raised the choppers to a hover. We did a pedal turn and hovered five to six miles down the length of the valley. The enemy forces must have been the worst shots in the world because they didn't wipe out any helos or even hit us.

Far enough down the valley the enemy fire ceased. In the distance I saw a searchlight shooting straight up, illuminating the clouds.

"That's where COL Moore is," the CPT said. "Our BDE HQ.

I called on COL Moore's frequency and told him I was inbound with his company and would be there in a few minutes.

"I can hear helicopters out there," he said.

"That's us," I said.

When we arrived I had my choppers land in a circle fifty to sixty feet apart in the firebase. I had the gunners and crew chiefs get out and join the infantry at the perimeter. I paced around the perimeter to make sure my guys were in position everywhere. I told my crews they were grunts now and needed to cycle sleep.

COL Moore appeared out of the background light.

"Who is in charge of this outfit?" he said.

"MAJ Gosney," one of my men said. "He's over there."

COL Moore came over and introduced himself.

"I'm Hal Moore," he said.

"Yes, sir," I said. "I know you."

"How in the world did you do this?" he said.

"I heard your Captain call for help, so I flew over and got them. We had to navigate out over the mountain and descend into the valley to get them."

"The 1st CAV Division told me they couldn't fly," he said.

"Maybe they didn't want to get airborne. This weather's a mess." I nodded toward my men. "My guys have been awake about twenty-four hours. I wonder if you have any coffee for them. They're on your perimeter right now."

He motioned to his SGM.

"Bring a big pot of coffee in a marmite container," he said, "and pass it around to his guys."

"I imagine your soldiers could use the hot java, too," I said.

He got that accomplished.

"Come on in," he said. "I want to talk to you."

I went into his command bunker. He was one of the nicest men I've met.

"Explain to me in greater detail," he said, "how you got here."

"We had been near Phu Bai operating on a mission for the Korean Tiger Division. We were coming back and I heard a guy on your command channel call for help."

"Yes," Moore said. "And I couldn't get a chopper out of the 1st CAV Division."

"We did it," I said. "We got them."

"Go on."

I nodded and finished the story.

"I'm amazed you rescued the company," he said. "You saved my backside. The VC tried to wipe out my unit but you stopped them."

"It's no big deal. When somebody's in need, we don't hesitate to help."

"Well, I respect you for that," he said.

"Yeah."

"How many of those guys are instrument rated?" He sipped his coffee.

"Two of us." I shifted in my seat. "Me and a CW4."

"All your other pilots aren't instrument rated?"

"No." I grinned. "I gave them a quick class while we waited to get the GCI vectoring. And they made it, smooth as silk. Some choppers waggled when they broke out of the soup, but they lined up straight as an arrow in the valley."

We stayed until daylight.

"I'm heading out," I told COL Moore. "Got to get over to Pleiku."

He shook my hand and congratulated me again for saving his men.

"You will receive a Silver Star," he said.

"I didn't do it for a medal," I said. "I did it to help a young CPT in trouble."

"I'm recommending you for the Silver Star."

"Okay, Colonel," I said.

He bid me farewell.

We mounted the choppers, came to a hover, got three abreast, and headed home.

Hal Moore called a few days later at Pleiku on our field phone, a small radio.

"I still want to thank you for bringing back the CPT and his men," he said.

"You've thanked me enough, COL," I said. "We've got another operation in about fifteen minutes, but I appreciate you calling."

Three weeks later I got a telex from Hal Moore reaffirming my name was going in for the Silver Star. In my reply, I stated I thought higher HQ would rescind the medal or disapprove the recommendation because I was only doing my job. As far as I know, COL Moore never submitted a recommendation, but our connection remained strong.

Many years later, I was watching TV with my eyes closed when a story about Hal Moore came on. Jean woke me and asked me if I knew him.

"Sure do," I said. I called him and we talked a little bit.

Years after that event, a similar happenstance occurred to one of my nephews, Matthew, an Auburn student. He and his buddy decided to mow lawns to make money. They determined to go into the affluent area in Auburn.

Matthew rang a doorbell and waited. A gentleman answered the door.

"Sir," he said, "I'm Matthew Gosney."

"Gosney?" the man said. "Is Bob Gosney any kin to you?"

"Yes, sir," Matthew said. "He's my uncle."

"He saved my tail in Vietnam." He introduced himself. "I'm Lieutenant General Hal Moore."

Matthew couldn't wait to get his telephone out of his pocket. He called me from Auburn in the yard he was about to mow.

"Uncle Bob," he said, "You won't believe what I found out."

"What's that?"

"Lieutenant General Hal Moore said you saved his tail one day in Vietnam."

"I don't know about that," I said, "but I sure helped out one of his officers."

"When I told him my name he asked if I was kin to Bob Gosney."

Matthew was just struck.

"Matthew, are you going to mow his yard?"

"Yeah, right now."

"Mow it good," I said. "He's a fine man."

Hal lived in Auburn while he was in the service. He had bought a home and he settled there when he retired. Soon after retirement, he got sick and over time passed away from Alzheimer's disease.

Loss of a Friend Flying a Grumman Mohawk

A friend of mine piloted a Grumman Mohawk in Vietnam. He was under GCI radar control because of the altitude of his aircraft. The controllers had him shoot over Pleiku. He called me on the radio and said he was on a secret mission. I wished him God-speed. That was the last I heard of him.

The controllers gave him headings and then lost him. There one second and gone the next. They couldn't raise him on the radio or visualize his radar signal.

He must have had an emergency or a missile shot at him. He was cruising about twenty thousand feet. At that altitude, the pilot has to wear a pressurized suit, a G-suit, in the cockpit.

The unit never recovered the body and found no wreckage. Strange.

I guess I was the last one he ever talked with before he was lost. He and I were in the same flight class together. His disappearance left an empty space inside for a while.

He loved piloting the Mohawk, a graceful and agile aircraft when the first aircraft debuted, which suited the operational missions he flew. Before modernization, the turbine engines made the plane a quick accelerator. If you set your flaps to the short field take off setting, revved your engines at full speed, and released the plane, you could be airborne in less than a hundred yards.

A North Vietnamese Woman Shoots and Kills a Lieutenant

I inserted one of my platoons on an ambush mission on the plains outside Pleiku. Navigating north from lower South Vietnam to its center, a flier crosses over mountains and the central highlands, which consist of level plains. We discovered an excellent place to set an ambush. I had a young West Point LT who was sharp as a tack. I talked to him about the ambush and he led his platoon in the operation.

"It could be like a turkey shoot," I said, "because you'll be shooting at such a distance the enemy can't hear you. Start at the back of the convoy and work to the front."

I was in radio contact with them, listening to the action. The LT called me.

"They're easy targets," he said, "like shooting ducks at a county fair."

A short time later we got a second call.

"Medevac," the soldier radioed. "We need a medevac."

I decided to fly in and get the casualty. It was the LT. They gut shot him and he didn't live long.

"Four Vietnamese women walked down a dike," his platoon sergeant said. "The LT called out 'Cease fire. There are women out there.' The LT went out to convince them to move away. One woman made a sudden move. She lifted her black gown, leveled an AK-47, and shot him on the spot.

"What did you do then?" I said.

"We killed them," he said. "Afterwards we ran out and grabbed the Lieutenant."

Our cover was busted now. The enemy knew where we were. I sent in a couple more choppers to retrieve the platoon and bring them out.

The young West Point officer wasn't going to shoot a female, so she shot him.

Alvin Langford

Vietnam could be a small place. I ran into a great friend of mine there, Alvin Langford. I've known Alvin a long time. We were at Texas A&M University the same time and he and I played together on the A&M football team. He was a U.S. Air Force KC-135 pilot. The aircrafts' primary usages were as strategic air-refueling planes.

My chopper forged into the area north of Qun Nhon and I heard a FAC (Forward Air Controller) calling in air strikes.

If you ever heard Alvin talk, you'd never lose him because he had such a peculiar voice. Alvin doesn't talk fast at all; he talks real slow. He was from north-side of Fort Worth, Texas.

I waited until Alvin completed the air strike. He was piloting an Army L-19 Birddog, a single-engine, high-wing aircraft made by Cessna. The Army used L-19s for observation and forward air controllers. When he had completed his mission, I keyed the microphone on the radio.

"Alvin Langford," I said.

Silence.

I keyed the microphone again.

"Alvin Langford."

"Who's calling Alvin Langford," the FAC said.

"One of your admirers." I had a sly grin on my face.

"Bob Gosney, what the heck are you doing?"

"I'm mighty close to you," I said. "Have you landed?"

"Yeah."

"I'll be at your air strip in no time." I change course and landed at his airfield. Alvin and I spent a while shaking hands, hugging, and looking like two old friends.

"Alvin, you look like a tramp," I said "Don't you have any jungle fatigues or something better to wear?"

"All I got, Bob, is what the Air Force gave me."

"Well, do you have any cold beer?" I said.

"I haven't seen a cold beer in months."

"What size shoe or boot do you wear?" I waved a hand at his feet.

"Eleven-and-a-half."

"What size pants?" I eased out my notepad and noted everything he told me. "Alvin, I'll be back here tomorrow at about two o'clock and I'll bring you your Santy Claus." We laughed.

I zipped back to Pleiku, the high ground. I had radioed ahead and the unit had a uniform ready when I arrived. We went through uniforms in Vietnam like water flowing through a streambed. My solders would be in the jungle for two weeks, then needed new clothes. I got the jungle fatigue pants, shirts, and brand new Army jungle boots that fit him.

"I need an ice chest," I said to one of my friends. "Whose got a big ice chest?"

"I do, of course," he said.

"I'll buy it from you." I pulled out some bills. "I have a buddy over here and he doesn't have any beer."

Believe it or not, we had an icemaker at Pleiku. The next day I filled the ice-chest full of beer and put ice all over the top. Canned beer. We didn't get bottled beer or kegs. I loaded the ice-chest, the clothes, and the boots in my chopper. We had a little store where you could buy chips, dips, and personal use items. I bought munchies and stowed them in the bag for Alvin. And away we went. My crew chief helped load the items on the chopper.

"Sir," the crew chief said, "can I have one of these beers?"

"Yes, but save it." I warned him. "Don't drink it while we're in the air. Do it when we RTB."

"Okay."

We settled on the helipad area. Alvin strode across the field from his tent.

"Alvin," I said, "I've got special things for you. Santy Clause came." I began unloading a big bag.

"These are your clothes and your boots. Here are the chips, dips, and other miscellaneous stuff." I paused. "Now this," I said as the crew chief unloaded the cooler, "this is what you're looking for. This is full of ice-cold beer."

"Aw," Alvin said, "my prayers have been answered."

He and I chuckled. He was elated. Two other Air Force pilots stood beside him.

"I'm sure you're going to save this beer for those guys," I said.

"I'll do that."

"Save this ice chest," I motioned toward the cooler, "because in a few days I'll get it, fill it, and bring you more beer."

"What can I pay you?" he reached for a wallet.

"You keep flying high in your L-19."

We laughed. He was supposed to be flying a KC-135, a much larger, four-engine aircraft, and here he was piloting a little single-engine, high-wing Cessna.

Pilot Ejects from F100 over Cambodia

One incident in Vietnam sticks out in my memory. I was leading the formation of Hueys as flight ops and heard a Mayday. My eyes scanned the sky and caught the glint of an F-100 Super Sabre starting to spin over Cambodia. A guy ejected out of the aircraft, chute popping open and jerking him to a gradual descent.

"Circle around here," I told my Exec. "We're going to get him."

I thundered into Cambodia. As he slammed into the ground, I landed forty yards from him. I didn't want to pull pitch too close, because that would reinflate his parachute and drag him away.

The pilot released his chute. I picked up and touched down right in front of him. He jumped in the back and I whirled around back into Vietnam.

"I'm dropping you at the Du Cho Special Forces Camp. They can arrange your RTB."

"That's fine." He patted my shoulder and reached over and shook the co-pilot's hand. We happened to be in the middle of a battle doing combat assaults. I dropped him off at Du Cho and came back to the fight.

"Did you get his name?" the co-pilot asked me.

"No, I didn't."

The last thing I wanted was his name. My main desire was to get him out of Cambodia.

He never tried to get in touch with me, I guess, but later, a guy visited our camp.

"He's rotated back to the States," the guy said. "He told me a helicopter called Clipper 6 had rescued him."

"Yes." I nodded. "My chopper is Clipper 6."

"I'll get the word to him."

In the end, I got a nice letter from him. He hadn't been shot down. He had a flame out he couldn't restart.

"I don't think the enemy took me out," he said in the letter.

When I saw the engine failure, he was a good six thousand feet in the air. But it doesn't take long for an F-100 to fall from that altitude to the ground. The plane is like a rock. The airframe has swept back wings with no aerodynamics to keep the plane aloft. Thrust alone keeps that aircraft in the air.

L-19 Forward Area Controller Pilot Shot Down

A FAC in an L-19 Birddog aircraft and I were both flying close to the Chu Pong Mountain. I listened to his radio transmissions as he reconned a valley. As the Air Force FAC, his reconnaissance mission was to check out where to call in airstrikes.

Suddenly, I saw his L-19 dive into the jungle. He crashed hard. I think he was shot down by small arms fire. I radioed him.

"Are you okay?" I said. I kept repeating the question. No answer.

I saw a lot of the North Vietnamese. They had dug in. He fired a .38 pistol—I'll never forget the sight. He dropped a couple of NVA.

"Get down there," I said to my gunships, "and lay fire support on the bad guys shooting at the pilot."

The gunships went on the attack. With mini-guns in the nose, they riddled the heck out of the NVA.

I glimpsed the Captain lying in a ditch.

"Dang, that doesn't look good," I thought.

I had a U.S. Army Air Cavalry scout platoon of soldiers with me on my chopper. I inserted them into the area and the gunships kept the North Vietnamese away as best they could—they killed a bunch of them, which allowed us to go in and rescue the Air Force pilot.

The scout platoon reached the pilot's position. He was dead, shot in several spots.

I landed and the scouts laid him in my chopper. The Air Cav platoon jumped on board and I took off.

The pilot was young guy. He might have flown fighters for a while followed by moving into an L-19.

I flew to Du Cho, which had a doctor. When I hit the dirt, I had the body carried to the doctor's location. He checked the Air Force officer.

"He's dead." The doctor shook his head.

"I figured," I said. "A lot of people were shooting at him."

I don't know who the young man was, but we tried to help him. The least we could do was repatriate the body. I didn't want the enemy to get to it.

Chu Pong Shooting Gallery

A number of interesting events...you might call them dangerous...occurred in Vietnam. As the 119th Aviation Company (Air Mobile) Commander, I took several choppers on a night recon excursion into the jungle right by the Chu Pong Mountain. The

Ia Drang valley near that mountain was where the famous battle took place in Hal Moore's book, *We Were Soldiers Once...and Young*. The Chu Pong Mountain is where the North Vietnamese gathered and then attacked.

If you drew an arc on a map through the Chu Pong, bulging toward Vietnam, that would be the Cambodian border. The border bisects the middle of the mountain, then heads south. Straight north, the border would still be in the center of the mountains. At night you could hover and observe trucks, headlights on, bumper-to-bumper, coming south on the old jungle road, the Ho Chi Minh Trail, on the Cambodian side. The trucks were North Vietnamese supply trucks. They had to keep their headlights on regular settings, otherwise they couldn't move fast.

I talked to my division commander about the situation. I had been reporting the issue for a while.

"I'm telling you, there are hundreds of them," I said. "It would be a shooting gallery. I could take my gun platoon and gunships and we could destroy at least fifty or sixty loaded trucks in one night."

"In Cambodia?" he said.

"Yes." There was a pause.

"No," he said. "But if those convoys were in South Vietnam, it would be okay."

I never got permission, but I set-up a night mission for a select group—five Cobra gun ships of mine. I had four slick gun ships that could fire twenty-four missiles off each side, totaling forty-eight missiles. My ship was command and control. I stayed at a pretty high elevation in the command chopper to see the trucks and orchestrate the attack. The unsuspecting drivers still had their headlights on. The attack was like picking off slow-moving targets at an amusement park.

I launched all gunships and started the attack at the convoy's rear, and finished at the front to stop them in place. The gunships destroyed a hundred loaded trucks. A lot of the trucks transported North Vietnamese soldiers and numerous secondary explosions meant some trucks moved ammo and explosives.

Attack finished, I called the gunships off.

"We're headed back to Pleiku." I radioed everyone at once.

We arrived at Pleiku before daylight.

"Don't mention where this took place," I told my troops.

I don't know the specific battle damage count, but flames from burning trucks lit the sky throughout the column and we came back with no ammunition left.

I'll bet the Corps Commander, a Lieutenant General (LTG), who might have been LTG Bruce Palmer, Jr, the II Field Force Commander from March to July 1967, knew all about it. The operation was too big for him not to know. He never said anything, but the Pentagon called him to the United States for a Secretary of Defense (SecDef) briefing.

Our highest headquarters for 1st Aviation Brigade was II Field Force, a three-star, corps-level command. My chain of command was the 119 Aviation Company, 52nd Combat Aviation Battalion, 17th Aviation Combat Group, up through the 1st Aviation Brigade, a one-star billet to the II Field Force, the three-star.

At the briefing to Robert S. McNamara, the SecDef, the Corps Commander made the fatal mistake of saying there were thousands of North Vietnamese soldiers in Cambodia.

"Are there any Vietnamese using Laos and Cambodia as a staging areas?" the SecDef said.

"Without a doubt, yes," the Corps Commander said.

"Wrong answer," McNamara said. "I want you to know there are no Vietnamese using those two countries. We have no information of foreign military soldiers there."

"I'm not going to lie to you, Mr. Secretary." The Lieutenant General was adamant. "At night you can see the lights of convoys coming down the mountain in Cambodia. They're extensive."

McNamara belittled him. "Are you saying you saw them?"

"I have seen them," the LTG said.

That three-star general never came back to Vietnam. All of the officers at our unit often wondered what happened to him. It could be because he was truthful. The Army published information which stated the senior staff was upset the Lieutenant General had the audacity to disagree with the Secretary of Defense. Senior Army officials must have fired him from his position because he never returned.

McNamara didn't want the truth; he wanted the LTG to say there were no North Vietnamese in Cambodia.

That's the last time I risked going to the Chu Pong Mountain to attack a convoy.

"My word," I thought, "if a three-star general gets fired because he says, 'Yes, they're over there, you can see the trucks,'" I wouldn't fare any better.

RESCUING POW JAMES NICHOLAS "NICK" ROWE

The Internet has many sites containing information about First Lieutenant Rowe, who was in Vietnam in 1963 as a Special Forces officer in the Mekong Delta. After several months in country, the Viet Cong captured Rowe and a few other soldiers as prisoners of war. They held him in the U Minh Forest, at the tip of South Vietnam, jailed in small bamboo cages. The VC moved him frequently.

I received a highly classified mission while at Pleiku. The scheduled operation was for the Mekong Delta, a long distance from my base. The Special Operating Group (SOG) Commander visited me. We ambled out on the airfield and he gave me a Top Secret mission. We were to conduct a raid in the Delta to rescue a prisoner of war (POW). His SOG team briefed me on the details.

The POW was James Nicholas "Nick" Rowe, a Green Beret Lieutenant. He was an amazing man. Tough as nails. He spent five years and two months without any activity in a cage in South Vietnam—1,888 days.

No one in my unit knew the entire mission except for the SOG LTC and me. I couldn't tell my troops any details. I told them we had to do a rehearsal.

In the highlands of Pleiku, we found terrain similar to the Mekong Delta area we would attack. There were two rivers in the highlands, like the fork in Junction, Texas, where the north and south Llano Rivers join. The targeted rivers in the Pleiku highlands joined together, forming one river, which worked as a substitute for the Mekong river. On the west side of the river was a soldier in a cage and on the other side the SOG directed us to place a blocking force. Then we were to descend with two rifle companies to complete the rescue.

We practiced the mission like a modern mission rehearsal exercise (MRE). One element of my unit would drop off soldiers on the west side of the river as the attack force. The other element would execute a simultaneous drop of troops as a blocking force on the east side of the river. The Mekong river was not deep; therefore, we had to have a blocking force to keep the enemy on the east side of the river.

I didn't tell the troops or pilots the full plan.

"Let's pretend we have a force of soldiers and we're going to drop them in here." I pointed to a map of the exercise area. "On the other side of the river we're gonna drop another group of soldiers to be a blocking force. We trained and rehearsed until we had it down pat.

We used Chinese Nungs as our soldiers for the operations. These Nungs were mercenaries and large groups of them worked with us. I had two hundred Nungs in the initial assault to free Nick Rowe and I had an American infantry company to land on the opposite side of the river as a blocking force.

The Chinese Nungs NCOs were from five different areas in the China mainland. The officer was always an American who could speak Vietnamese. The officer had a company of 150 Nungs. They didn't use American weapons. The Nungs worked hard at being soldiers who could withstand rugged terrain and close combat. They got paid $5-a-day while in base camp. If they went into the bush, they received $10-a-day. When you asked for two hundred volunteers to go to the bush, every one of them wanted to go because of the double pay.

The SOG sent five C-130's to move my men except for the pilots and a crew chief for each chopper. We took the choppers to a place called Tay Ninh in the Mekong Delta. The C-130's met us there. The Chinese Nungs company commanders, sergeants, and troops had been pre-positioned at Tay Ninh. Once we were all in place, we briefed the plan to everybody, the plan to rescue the American POW, Nick Rowe.

We had planned and rehearsed down to the tiniest details. We had four hundred Nungs. We put fifty Nungs in the blocking force with the American units. The rest of the Nungs were in the assault force.

Everything we rehearsed for our forces went as planned except for one glitch: our intelligence didn't reveal the Viet Cong 9th Division was operating in the area. The 9th Division was composed of Irregular North Vietnamese forces.

We made the assault, but landed right in the middle of the VC Division. That's how bad the intelligence was in Vietnam. The assault force was in combat the minute we dropped them from the choppers. We inserted the blocking force unopposed. I saw we were up against a formidable enemy.

All hell broke loose. The plan was to give the assault force thirty minutes, then I would call in my lift ships to retrieve the Chinese Nungs. The Nungs should have released the POW by that time. However, I didn't see the operation happening that fast with the incredible resistance we met.

The Chinese Nungs were fighting little devils and went into combat like a hot knife into butter. They piled out of those choppers, shooting as they came out. Their quick action killed everybody trying to hinder us. When we finished, we had routed two battalions of the 9th Division, which would have been about a thousand to twelve hundred men.

The worst of the fighting settled down in twenty minutes, but the American company commander warned us off.

"Don't come in," he radioed. "We are fully engaged."

In about thirty minutes the firefight had died out. The Nungs had routed the Viet Cong and killed most of them. Our comparative losses from the firefight were miniscule—two Chinese Nungs.

I landed. We went into the brush and I found the bamboo cage. Carved into the wood were the dates he had stayed in the cage. His captors had moved him in the night. The cage was twice as big as a small chest of drawers: six feet long by four feet high by three feet deep. He could at least sit if he wanted. He stayed in the cage most of the time he was a prisoner. Rowe had carved numbers like 1, 5, 4, meaning Jan 5th, 1964 or whatever year it was. He carved in every cage, which gave our intelligence officers good clues as to where the VC had been moving him. But his captors had moved him under cover of darkness early in the morning before we executed the assault.

The mission was a great disappointment. The entire operation seemed like a lot of preparation and fighting for nothing. I found out when they did rescue Nick Rowe. After he escaped on December 31, 1968, he had been the longest person in captivity in South Vietnam. Later, the Viet Cong released another officer in 1973, Floyd James Thompson, ten days short of nine years in captivity.

The other people captured with Rowe didn't have the will to live like he did. A nurse taken captive with him died, even though she was strong. Since she was a woman, the VC didn't give her anything to eat. If she caught bugs or something, she would eat the bugs. Her captors let her eat what she trapped in the bamboo cage, like cockroaches or other bugs. The VC starved her to death instead of using her nursing skills. When she died, the Viet Cong didn't even bury her. They drug her into the brush and left her body.

A couple of other American soldiers died because they gave up. The VC kept prisoners naked and the mosquitoes ate them alive.

Six months later, a SOG team was in the Delta again. They established a small firebase of four Cobra helicopter gun ships and two Loach helicopters. The whole team's mission was to find the Viet Cong in order to employ soldiers to engage them. Near the end of December, a Loach team in their scout helicopter worked recon in the Mekong Delta area. A Loach is a small helicopter seating two men. The Loach team went out, found targets, and called in gun ships to have them destroy their find.

For this operation, the Loach team also had two UH-1 Huey gunships escorting them five or six kilometers from where they made the assault. They reported a guy running down a rice dike and into the rice paddies. The man didn't have any clothes on. They prepared to engage their weapons.

"No," the Loach pilot said, "Not if he looks like an American."

The team landed and the man was Nick Rowe. He was buck-naked. He was skin and bones because the Vietnamese fed him next to nothing. He had broken out of the cage when choppers made a lot of racket overhead. They rescued him in the little chopper and flew him to a UH-1, which flew him back. He was free after five years and two months in the jungle.

"Take him to Long Binh hospital," the SOG commander said when the team on the ground found out the man's identity. "I'll meet him there."

The team delivered him to the medical facility, located near Saigon. It took him months to regain his strength even though he was in amazing physical shape for a POW. He had lost sixty pounds and was skin and bones. The doctors treated him for malaria and a variety of symptoms from malnutrition and disease. He looked like one giant welt of mosquito bites.

While he was recovering, the SOG visited him to talk about our operation several months back. He knew what we were trying to do with the operation. The Viet Cong had moved him a mere five or six hundred yards away from where we thought his captors had put the cage. He heard the battle and convinced himself we were coming to get him.

The Army gave him a retroactive promotion to Major, the rank he would have held had he been in the regular promotion system. He rejoined his unit stateside.

The sad thing about Rowe's story came a few years later. He became a full Colonel assigned as the Attaché to the Philippines Embassy in Manila. One morning he drove to work in an armored car and the New People's Army (NPA) assassinated him. Rowe was doing intelligence work against the NPA and knew he was number two or three on their list for assassination. After everything the man had been through, I think that's a tragic ending to his life.

He was born in McAllen, Texas, which remained his home. He published a book in 1971, Five Years to Freedom, to detail his entire experience.

CHAPTER 5

Preparing for My Second Vietnam Tour

When I returned to the Continental United States (CONUS) from my Vietnam tour, I immersed myself in my project officer duties in the Employment Branch, Systems-Effectiveness Analysis Division of the U.S. Army Combat Development Command Aviation Agency in Fort Rucker, Alabama. I directed and assisted in developing scenarios for the Utility Tactical Transport Aircraft System (UTTAS) Study.

These two years of working on aviation systems increased my knowledge and preparation for a second tour in Vietnam, this time as a Battalion Commander. However, the mission of first importance was moving my family from Fort Rucker to Oahu, Hawaii, my home base of assignment. I would deploy from Hawaii to Vietnam soon after we arrived on the island. As any Army officer worth his salt can tell you, ensuring your loved ones are settled prior to deployment is a key factor in maintaining a clear focus in the combat zone. I didn't have to worry about Jean and the kids—she was an expert Army wife and could handle whatever occurred while I was away.

Traveling the SS Lurline to Hawaii

The Army assigned me to the 25th Infantry Division as the Commanding Officer of the 25th Aviation Battalion. But Molly, one of my daughters, couldn't travel by air.

Molly had open-heart surgery when she was four or five years old at Brooks Army Medical Center. The surgeon who operated on Molly had been trained under Dr. Michael Debakey, a world-renowned Lebanese-American cardiac surgeon. We had great confidence in Molly's surgeon; however, during the surgery, one of Molly's lungs deflated.

"Don't worry about it," the surgeon said. "The lung will come back. It might take a year, but it will reinflate."

Therefore, even after several years, the doctors still restricted our family travel, meaning we couldn't fly to Hawaii.

I took the issue to the Fort Hood Transportation office.

"I've got a problem," I said. "I need to go to Hawaii by sea because my daughter's doctor said she should not travel by air until he examines her a couple of years from now." I explained the background.

"Let me see," the transportation clerk said. "Would you like to go on the SS Lurline?"

"Sounds like a cruise ship," I thought to myself. And it was.

"I thought you were going to put me and my family on an old troop ship."

"No," she said. "You are going in style."

We sailed in luxurious comfort. The trip was a treat, considering my final destination was Vietnam and I had uprooted Jean and the kids to move.

The kids loved Hawaii the entire time we were there and didn't want to leave.

Hawaii Quarters and the Ticket on the Pali

The world can seem like a small place when you are reassigned in the military. Arriving at the Honolulu Airport in Hawaii, I anticipated seeing Rabbit and Jenny Hare and their family, great friends from our college days at Texas A&M University.

The Hare's lived in Kailua, a town on the opposite side of the island from Honolulu. Rabbit and Jenny helped me settle Jean and the kids by finding a house for us to rent, not far from where they lived. Rabbit's Air Force assignment was at Hickam Air Force Base, a few miles outside Honolulu. The drive across the island to Kailua was short on the Pali Highway. Any driving we did to Hickam, the Honolulu airport, or other military bases seemed a half hour or less.

The house Rabbit and Jenny found for us was good. We lived

near the Pali Mountain Pass, the Pali Highway route to cross the island. Descending from the Pali pass, a ridge extended off of the highway. Our house sat on that ridge. We could gaze up at the Pali Mountains, luscious tropical green and beautiful. Our house didn't have air conditioning or heat, but the temperature never got too warm or cold. After I made sure Jean and the family was settled in, my time with them disappeared.

I had two weeks before I had to catch a flight to Vietnam. The U.S. Army provided Jean and I Army cots because our household goods were on another ship in transit. The kids had pallets on the floor. Our next-door neighbors were the Noweskas, a native Hawaiian couple with twelve kids. They were the best mannered children I have met in my life. And the Noweskas adopted Jean and our kids.

Right before I had to leave for Vietnam, I was driving back home on the Pali Highway. I had gone over to Honolulu to ensure the flight to Vietnam would take off on time and rented a car for Jean and the kids to have until our car arrived from CONUS.

I drove off the Pali Mountain and got stopped for speeding. The policeman wrote the ticket and gave it to me. I drove a direct route to the police station in Kailua. A big Hawaiian desk sergeant watched me enter the building.

"What can I do for you, bro?" he said.

"I got a speeding ticket." I handed him the paper. "I have to pay for it now because I'm leaving in the morning and won't be back here for a year."

The sergeant glanced at it.

"Has he turned the ticket in yet?" I asked.

"We're modern here. The policeman writes it out and forwards the electrons via automatic relay. Seven dollars and fifty cents." He handed the ticket back.

I pulled out all I had—a twenty-dollar bill.

"Oh, bro." He smiled. "I can't cash that. I don't have any change."

"You can't change a twenty?"

"No. The ticket's only seven dollars and fifty cents."

"It's the smallest bill I've got. And I'm leaving in the morning." I knew I had a lot to do to finish getting ready.

"No problem, bro." He smiled and made the Hawaiian shaka hand gesture. To make the shaka, extend your thumb and little finger keeping the rest of the fingers curled in your palm, then rotate your wrist left and right. "I'll see you when you get back."

I came back sooner. Vietnam R&Rs (Rest and Recuperation Leave) lasted seven days. Not much time. But on my first R&R back to Hawaii, I made a bee-line to the police station and paid the desk sergeant seven dollars and fifty cents.

"I thought you'd be gone for a year," the sergeant said. "I still remember you."

"No," I said. "I owed you the money and I don't forget to take care of my debts, even if they're as small as seven dollars and fifty cents."

What amazed me was that he remembered my name.

CHAPTER 6
VIETNAM II

My second tour to Vietnam in 1969, I commanded the 25th Aviation Battalion, 25th Infantry Division. This was a reinforced battalion consisting of three lift companies, two gun companies, and an infantry company. We flew many intricate missions in South Vietnam and in Cambodia.

In spite of the intense combat, there were lighter moments. Rabbit Hare, my Air Force fighter pilot buddy, would send me a penny post card while I was in Vietnam. "Send more booze," he would write.

Military returning from Vietnam stopped in Guam, a duty-free port, to refuel. Bottles of the finest scotch cost four or five dollars with no taxes or import fees. I had several officers or soldiers I knew buy a few bottles for my great friend, Rabbit.

The mission in Vietnam remained a non-stop experience, seven days a week. Now I averaged about twelve hours flying a day.

Though the U.S. military presence in Vietnam had expanded, new soldiers didn't know enough to avoid their own mistakes. Training from experienced leaders was paramount. When soldiers first hit the dirt in Vietnam, their NCOs and officers had to eradicate fatal habits, like wearing aftershave, cologne and deodorant. The VC could sniff out a new arrival in minutes. We trained soldiers in essential techniques to survive their tour. I had the opportunity to teach my younger brother a few things to keep him from meeting an early demise.

Young eighteen-year-old pilots coming to Vietnam faced one major internal problem for operational flying—they had no fear. And a few pilots suffered for their faulty decisions.

Looping a Huey to Create a Training Prop

One of my young pilots thought he could loop a Huey. A Huey helicopter is not designed to perform loops like a fixed wing aircraft. This youngster got the chopper halfway around his loop. As he strained to pull out of the maneuver, he couldn't stop the helicopter. He tore into the ground, broke his leg, and got cut pretty bad. His co-pilot sustained severe injuries. The doctors sent him to the major hospital in Saigon, where they fixed his leg, patched a couple of broken ribs, and took care of the other medical issues. The pilot came back with a big cast on his leg and a cast on his broken arm reaching all the way to his shoulder.

I used him as a training prop. When I'd get a bunch of new, young warrant officers, I'd call them into a briefing room with this pilot in a cast.

"See this guy?" I said. "He's stupid. He thought he could loop a Huey. It's impossible to loop the Huey without crashing. The bad guy's not going to kill you. Your going to kill yourself if you do stupid things like this."

Warrants shifted in their seats, some eyes glanced away, and others raised their eyebrows. My speech and the injured pilot made an impression.

"This guy's lucky." I'd lock onto their eyes. "He survived crashing the chopper nose down. But I'll tell you, he's staying here and I'm going to make a helicopter pilot out of him or he'll kill himself trying."

I used him as a sample of stupidity until he was well. When his doctor cleared him, I had his company commander put him in a chopper. He became a co-pilot with a good pilot beside him, training him.

In the long run, he made a good pilot.

Condemned Steak and Lobster Dinner

On my second tour, my younger brother, Gary, and I were in Vietnam at the same time.

"Now, Bobbie," my mother told me, "you take good care of Gary-boy."

“Okay, mama,” I said. “I’ll do that.”

Gary arrived in country before I did. I was based at Cu Chi, on the outskirts of Saigon, not far from Gary’s quarters.

Gary was the Saigon Veterinarian and lived in designated quarters: a large, fenced French villa with window air conditioning, which was most unusual.

I flew to his house one day for dinner from Cu Chi. Gary had invited a large group of people and served a steak and lobster dinner, also unique in Vietnam. A master-cook must have prepared the food, as it was excellent. I caught sight of someone in a chef’s uniform later in the evening.

“Gary,” I said after dinner, “how in the world did you find steak and lobster?”

“I just went over and condemned it.” He shrugged, a smile plastered on his face.

“You condemned it?” I lifted an eyebrow.

“Yes.” His smile widened. “I go to the screening station where the food arrives. I inspect it and say, ‘Those steaks—I need to check them for Salmonella.’”

He had condemned the lobsters and steak we had for dinner. General Creighton Abrams, Commander of MACV (Military Assistance Command – Vietnam) was a big steak eater. The logisticians shipped special steaks specifically for Creighton Abrams.

We probably enjoyed a bit of food meant for him. Delicious.

The Pig and the Thermometer

Gary traveled many places to perform his veterinarian duties. He left Saigon to go into the field, staying in a tent. An order of nuns had animals Gary treated. They didn't have a lot of livestock—most were pigs, chickens, and goats.

Gary monitored, checked, and treated the animals. While he was checking the temperature on a pig with a rectal thermometer, the enemy attacked, weapons blazing. Gary raced into the barn. One of the nuns got him to run into their place, which had a basement to get away from the gunfire.

Gary never revisited the pig. He headed straight back to his tent.

Gary laughs about this story. "Somewhere in South Vietnam, there was a pig running around with a thermometer up his you-know-what."

Gary Gosney's Reassignment from Saigon to Cu Chi

Two days after the pig incident and weeks after the mouthwatering steak and lobster dinner, I flew to the Mekong Delta to pickup an RVN (Republic of Vietnam) COL for a recon flight because we were planning a mission there. I finished the requirement and dropped the COL off. Next, I decided to stop and talk with the head of the RVN military in the region.

I set down where I was to meet the officer. I pulled pitch to settle in and my rotor wash blew down a tent twenty feet from my chopper. That time of the year was the rainy or muddy season.

Violent motions shook the tent, like a big pig trying to escape.

All of the sudden my brother crawled out from under the mud-caked tent and saluted me with two fingers, the middle one on each hand. Gary didn't know I was the pilot and I didn't realize he was in the tent until I blew it down.

I jumped out of my chopper, stomped his way, and lectured him.

"What in the world are you doing this far south? This is a long way from Saigon."

"I drove from Saigon," he said.

I got through chewing his butt.

"You get into this helicopter," I said. "I'm taking you back to Saigon."

"What do I do with my jeep?"

"Leave it," I said. "Somebody will use it." I loaded us back on the chopper and carried him back to the Saigon heliport at Bien Hoa. Then I went straight out to Cu Chi to see my Division Commander.

"General," I said, "I have a kid brother over in Saigon and I found him south of Saigon in the Mekong Delta. He had driven a jeep there."

"Are you sure he's smart?" he said.

"Yes, he's a brilliant guy." I nodded. "But he's not real smart about taking you driving out in the country."

"I'll get him out of there."

I gave him Gary's Social Security number and he arranged

his reassignment. I called Gary.

"Get packed," I said. "I'm coming to get you."

I picked up Gary at eight o'clock the next morning. I loaded him in the chopper with his veterinarian accoutrements he had brought. We navigated to Cu Chi where he became the Cu Chi vet.

Fill those Sandbags, Gary

We had about four-hundred-and-fifty war dogs in our area. We needed a vet. The vet's hooch was right on the perimeter. From Gary's tent there was twenty yards separating him from the concertina wire and minefield around the fence.

I told Gary one last bit of advice before I left to fly a mission.

"Gary, you haven't got a sandbag out here. Get your men and have them start filling sandbags around this daggone tent."

My Command Sergeant Major (CSM) went out to the tent first light the next morning. He took one or two of the Vietnamese hooch maids to start filling sandbags for him. That night we had a penetration on Gary's side of the perimeter—lots of shooting in that direction. When it calmed down, the CSM got busy.

"I'm going out to check on the Captain," he said.

The CSM drove out to the vet's tent at four o'clock in the morning.

"Boy," Gary said, "why are they shooting out here all the time?"

"Well, CPT," the CSM said, "that'll happen every now and then."

I drove out later and looked at Gary's tent. You could see daylight through a number of bullet holes. I saw one near his cot. I took a piece of long bamboo rod and stuck it through the hole and out the other side.

"Gary, " I said, "can you lay in there now?"

"No," he said. "I can't."

I pulled the rod out.

"Now get in." I examined the hole with him in the cot. The bullet would have taken his nose off.

"Now believe me." I leaned forward into his face. "Fill sand bags."

He took the message to heart. Gary was there nine more months and he sandbagged the whole time. I think his sandbag wall could have taken a ten KT (Kiloton) nuclear burst.

Major General Harris W. Hollis' Table

My division commander idolized my brother, Gary. He can tell the funniest jokes. I'd come in from the bush and have a meal at the General's Mess. Commanders had a table of their own.

Gary sat with General Harris Hollis on his right. That was his assigned place to eat. The General's table was round and seated lots of folks. The table had a big Lazy Susan in the middle of it. Most of the condiments, flatware, and food were on it. Under the table at each seat was a button. If you wanted ketchup or hot sauce, you pressed the button and the Lazy Susan rotated around. You released the button to shut it off at the right time and get what you wanted.

General Hollis had an override. His favorite thing to do when somebody ate at his table for the first time was to wait until they had rotated the table around to get something, then he punched his override to swing it around the other way.

He did that to me when I had eaten with him. I knew what he was doing. The trick created a little joviality to keep up our spirits.

General Hollis also extended me in Vietnam.

"Bob will go home when I go home," he said.

And I did. I commanded a battalion in combat for eighteen months instead of twelve.

MG Hollis was a great commander.

Landing an Infantry Brigade While Fighting Through AA Guns

For one combat assault mission, I had my choppers out searching for another LZ. We planned to land part of a 25th Infantry Division brigade—two battalions and the brigade headquarters.

The gun ships maneuvered in figure eights, shooting in the perimeters surrounding the LZ. While this was happening, I saw the tail boom fly off a gun ship and the chopper tumbled into the jungle.

"Good grief, what's that?" I said.

Another chopper on the opposite side got hit. His tail boom didn't come off, but the ship started to lose control. The enemy shot a hail of lead at it. The chopper spun into the jungle. Then another went down.

I called on the command channel, telling my Executive Officer (XO) our crew was going to find out what happened and now he was leading the landing.

"Turn around," I said. "Go back and wait until I tell you to bring them in here." I had four A1E Skyraider bombers on station. I radioed them.

"I'm marking the LZ we're going to use," I said. "I'll fire smoke into the LZ to indicate where we need heavy strikes in the surrounding area.

I descended to an altitude of fifty feet, treetop level, with my throttle wide open. I raced in at 140 knots, the top speed for a Huey.

I had the smoke rockets, since the command ship fires them. I fired the smoke, scanning in front of me and—BAM—eight hut roofs opened, each with an anti-aircraft (AA) gun plus one rangefinder that popped up. The AA guns blazed away at my helo, bullets raking the sky.

I pushed the trigger to fire the smoke rockets and did the fastest one-eighty (180-degree turn) I'd ever done in my life. I banked the aircraft on its side and bullets zipped through the chopper. The doors were open on both sides. When I came around the bullets went straight through and didn't hit a thing. I was shocked no one got hurt. The tracers were as big as basketballs. That's the closest I have ever come to being blown out of the sky. Those guns had picked off the three helicopters earlier. I gained altitude to watch the A1E's execute their mission.

The A1E's spotted the smoke and screamed in their vertical dives, dropping tons of unguided bombs to make their strikes.

One of the A1E pilots that day was a classmate of mine from

Texas A&M. I radioed him and asked if he was Ronald.

"Yeah," he said. "My name is Ronald. Who's this?"

"My name is Bob," I said.

"Oh, my goodness," he said.

"Give that daggone area a good bombing on all sides, even in the jungles on each flank."

They dropped massive amounts of bombs. Their team called and got four more A1Es to come on station. They operated out of Pleiku. The second group of A1E's bombed the thunder out of the jungle. They did a magnificent job.

I had one rifle company for my outfit. I deployed the unit after the bombing ended to take out everything they could find. I told them to eliminate the machine guns and rangefinder. When they finished, they loaded everything in a Chinook. They found eight Russian 51-caliber anti-aircraft machine guns and a Russian Range-finder. Those air defense guns fired with great accuracy and a rapid volume of fire.

That was exciting. The only time I almost jumped out of my skin. I had never turned a Huey around at such a steep bank angle, an uncommon maneuver for Hueys.

On one of the downed gunships, the gunner who sat in the nose of the Cobra survived. The Cobra helicopter has the gunner sitting in the front and the pilot sits higher behind him. The gunner fires the rockets and all the other weapons. He had survived the chopper's tumbling into the jungle after its tail boom was shot off.

He crawled through a North Vietnamese outfit and beyond them, half-submerged in a creek. By sheer luck the stream bent and came right to our location. The gunner was a young man; he looked like a teenager, one that was pretty banged up.

When my guys saw him, they moved out, got him, and brought him back to our unit. The medics worked on him for a while. He had cuts and bruises all over him. We airlifted him to the hospital at Pleiku. The doctors had to place him in the Long Binh hospital for a couple of weeks. They don't keep someone long without a major illness or concern.

When he got better and returned to the unit, he had two months of his Vietnam tour remaining. I sent him home early.

"Son," I said, "a cat's got nine lives, but we humans have got a mere two or three. You used all of yours at one time. Therefore, you're going home."

He was tickled to death. He lived in Missoula, Montana, the location of the home of the United States fire-jumpers.

We lost one pilot of the three downed helos. We recovered his body. The whole crew of the chopper with the shot-off tail boom survived.

Attack Into Cambodia

As the 25th Infantry Division (ID) Aviation Battalion Commander, all aviation in the division fell under my battalion. My reinforced battalion had close to twenty-five hundred troops. Our division commander, General Hollis, named us Battalion Task Force Gosney.

Our mission was to engage the enemy wherever we found them.

MG Hollis was an excellent General. He gave me free reign to do my mission and command my unit.

When Nixon was the United States President, the 25th Infantry Division was ordered to operate both inside and outside of Vietnam. The President talked a big game, telling the world we were going to cut off Cambodian supply lines to North Vietnamese.

"We are now operating in Cambodia," the President said.

When he made that statement, we didn't have any fighting men in Cambodia. I was stationed at Cu Chi, which is near the Cambodian border.

The Division Commander called me.

"I want you to lead the attack into Cambodia," he said. "The objective is to seek and destroy. Uncover caches of weapons and supplies, and burn them. If there are tons of rice, dump it in the river."

My role was to lead the southern half of the attack of a huge wing-like area on the map. The mission task force was an Infantry Battalion, an Air Cavalry Troop, and an Air Mobile Battalion. I was to head south, capture what I could, then meet the U.S. force closing down from the north. Loading out choppers, we were to spend a day having Chinooks fly the spoils back into the division. The plan and execution were superb. We just kept pressing into the enemy.

I spent most of my time on the ground with the infantry in the early going because it was hard to command that action from the air. When we hit roadblocks, I left a company to engage and lifted another to the other side, executing local pincer operations.

But we couldn't transport all the tons of rice back, because the weight was too much for our lift capabilities. And we couldn't let the enemy have it.

I had the Chinooks follow us, haul thousands of bags of rice in sling-loads, and dump the sling-loads into the Han River, to ruin the rice. We uncovered more supplies and equipment at our rapid pace, moving at breakneck speed. The battle rhythm was to make a decisive advance, engage and destroy, then sprint to the next objective, engage and destroy, and keep the cycle going.

I got on the secure radio with the Division Commander.

"Bob, you're not any farther than ten miles into Cambodia, are you?" the Commanding General (CG) said.

"Yes, I guess I am."

"The radio's secure," he said. "What's your exact location?"

"I'm on the outskirts of Duh Non Kori. I can see Phnom Penh from where I'm sitting." Phnom Penh, the capital of Cambodia, is about eighty miles from the Vietnam border. "I'm about a mile from the capital." I followed the CG's orders—tear them apart.

"That's not ten miles inside the border is it?"

"No, General. I'll make sure I don't go further than ten miles. I'll fly a hasty withdrawal."

"Pull back."

"Okay, sir."

"Have you had much success?" the CG said.

"Oh, heck, yeah. We've been killing them right and left and haven't lost anybody."

President Nixon had changed his mind in the middle of our operation.

"We won't go more than ten miles into Cambodia," he had told the senior commanders and the press.

I did a hasty withdrawal back to the prescribed limit. The restriction gave us time to haul lots of rice. We found five VW Volkswagens preserved in Cosmoline, a protective coating on brand new autos to keep them from rusting. It's like a brown gel you can wipe off. I had my lift assets take two VWs back to our battalion and had the other three dropped in the river where the Cosmoline and the cars would disintegrate. We dumped the rest of the weapons and equipment we found in the river.

When we got back to the base camp, one of my soldiers asked a question.

"What are we going to do with the Volkswagens, Colonel?"

"Get the daggone grease off and drive them," I said.

The soldiers got the Cosmoline off and they drove the Volkswagens around our camp. They were brand new and were both painted black. The soldiers had a lot of fun tooling around in the VWs, driving the sentries to our outpost and bringing them back. The ride was much better than a jeep.

Bob Hope and Neil Armstrong

While I was at Cu Chi, Bob Hope came to entertain the troops at Christmas-time. I had been there for six months as the Task Force Commander. Bob Hope came to our location twice while I was there. The first time he brought Neil Armstrong and a few pretty girls.

Neil Armstrong was the most significant man to me. I was a

bit awestruck. Six months ago he had been walking on the moon. But he was the most modest, introverted guy you could meet. He was almost embarrassed to say he was the first man to walk on the moon. He enjoyed the tour with all the military people there.

Landing Zones—Loss of Gunner

In Vietnam, we flew into jungle LZs and executed what was called Arc Light. B-52 bombers were involved. On one of these operations, we had five B-52s; however, events didn't unfold with the smoothness we desired.

"Synchronize your watches," the Air Force Liaison Officer (LNO) in my chopper said. "The bombers will be on station ready to drop in ten minutes. I'll relay the countdown from the bombers to you."

I lined up my choppers, twenty-four of them in V's of three each. I flew above them and made a mental count of the time as the B-52 pilots counted down for me.

"Sixty seconds to impact," the LNO said. Then he counted out loud.

I had my choppers start in. The plan called for our helicopters to start in a position a minute out. At thirty seconds remaining we would begin our approach. After the bombs had hit, the helos could proceed another twenty seconds to the LZ and hover to let the troops jump out.

Arc Light bombing was a sight to see as shock waves rocked the jungle and debris exploded in massive quantities. When the count got to zero, the whole world flew apart in front of us. Awesome. The B-52s dropped five hundred and seven hundred-

and-fifty pound bombs—lots of them.

I called my Exec leading the first flight.

"Take them in," I said.

Smoking stubble and stumps still remained from the bombing. My choppers hovered and the soldiers jumped out and ran. I had the Battalion Commander of the unit in my chopper. I swooped in to let him jump.

We had already put a battalion (BN) of infantry on the ground.

I don't know how the Commander survived the jump, because a daggone Vietnamese soldier shot my gunner, who was leaning forward returning fire. The bullet went into his neck, then into his heart and killed him. He's the only person I had killed on my aircraft.

I was at a hover and my gunner, sitting on the backside, fell back. I didn't see it, but the crew chief on the opposite side let me know.

"I think he's been hit," the crew chief said. "Henry's been hit."

"Look back," I said to my co-pilot.

"Did he fall out?" I asked my crew chief.

"No," came the reply. "He's here, but I think he's dead."

"Can you see blood?" I said.

"No, but there's a little trickle here on his neck."

"Check him out." I spoke into my mic. "See if he's alright."

"I think he's dead," my crew chief said.

On the way back, we dropped his body off at a supporting Graves Registration unit in Dak To. When we landed, I ran in and the unit took the gunner inside. The doctor from the Graves Registration unit listened for a heartbeat and breathing, then made his determination.

"No," the doc said. "He's dead."

"Of all the garbage I've had to fly through," I said, "that was the only time I've had a crew member shot and killed."

I wondered how any Vietnamese soldier could shoot at us after being in the vicinity of those huge bomb explosions. Where was he hiding? He must have been in a deep, deep hole.

My Senior Driver and the Air Conditioner Puzzle

Like most battalion commanders, I had a driver. He was a young man from New Baldwin, New York.

By sheer coincidence, my old Field Artillery Battalion Commander, Hugh Garden Brown, came from that town. When COL Brown retired, he returned with his wife to New Baldwin.

The Brown's read in the newspaper one day about a man from New Baldwin, New York, who was the driver for the senior Battalion Commander, Lieutenant Colonel Gosney, while he was in Vietnam.

"By golly," COL Brown said, "here is a man who went back to Vietnam and now says he is the 'senior driver' of the Senior Battalion Commander."

COL Brown informed me and we both chuckled about it.

However, the reason this man became my driver was because he had been shot twice for something stupid. He walked in from a sentry outpost and forgot the password. Then he ran back out to the sentry post and somebody shot him. Twice his own troops shot him in the butt.

I decided to make him my driver. The shots in his buttocks weren't too bad.

During the Vietnam monsoon season, I had a meeting to attend, but my tent was about a hundred yards from Division CP (Command Post).

"Drive me to the Division CP," I told my driver.

The rain fell like a waterfall—a real downpour. We even put the top over the jeep. The driver got me to the CP where an overhang protected me as I jumped from the jeep and went into the meeting.

"Wait for me." I turned back toward him. "When I come out, I'll holler for you."

Wouldn't you know it, he drove back to my Battalion CP.

The rain dumped sheets of water. I came out from the meeting and heck, there wasn't a jeep in sight. I thought, "Where in the world did that little idiot go to?"

I walked and got soaked to the bone in seconds. I stomped into my headquarters tent and there with my Command Sergeant Major was the driver, sitting and jawing away.

I was ticked-off to the max.

"Where did you go?"

"You told me to wait for you and you'd tell me when you're ready for me," he said.

"I didn't say right here."

The Sergeant Major started ripping the driver apart.

"Wait a minute, Sergeant Major," I said. "This misunderstanding could have been my fault because I didn't point to him and say, 'Stay right here. Right here. And you don't move.'"

My contribution made it even worse. The Sergeant Major was livid, he was so angry with him.

Later, I went on an R&R trip to Kuala Lumpur, the only R&R location open then.

The headquarters asked who wanted to go.

"I'll go," I said.

The trip was with a bunch of Australians. We were gone five days.

When I returned, I headed straight for my hooch, which was made out of old ammo crates and had a canvas top on it. I had an air conditioner. My supply officer obtained it somewhere. I mounted the unit in a crate and it kept my hooch cool.

Stepping inside, I stopped. The AC was now in five hundred parts.

I looked around, taking it all in. My driver, Mike, was sitting there.

"Mike," I said, "Do you mind if I ask what in the heck you did to my air conditioner?"

"Sir," he said. "I was fixing it, but I can't figure out how to put it back together."

"Then you didn't fix it, did you?"

"No sir, I've been trying, trying hard to get it back together."

"Oh, good grief," I said.

The one air conditioner in the whole division was sitting in my tent, scattered all over the ground. I felt upset, but I liked this kid. He wasn't smart at all. But I liked him because he was sincere.

"Do me a favor," I said in a quiet voice. "Go find a daggone garbage can. Put all these parts in the garbage can and throw it in the garbage pit."

" I can get it back together, sir." Hope shone in his eyes.

"I don't want it all back together," I said. "You could never put this AC unit back together. I'll have to get another one."

He hauled the mangled mess out of my hooch and I don't know what he did with it. But he hauled all the parts out.

The tools he had to work on the air conditioner were a screwdriver and a pair of pliers.

"How did you do that much damage with one screwdriver and a pair of pliers?" I said.

I didn't expect an answer and he didn't give one.

His tinkering took care of the one functioning air conditioning system in our unit.

THE PILOT IN THE RED SCARF AND THE PRISONERS TIED TO SKIDS

I had an officer piloting a Cobra gunship, a High School to Flight School program Cobra pilot. His father was a neurosurgeon. The young man joined right out of high school, came in the Army as a Warrant Officer (W-1), which was the lowest grade of a warrant officer. The program trained him well and he qualified in the Cobra. Right afterward, he was sent to join my unit. The boy was a tall, good-looking kid, but he was fearless on operational flights.

In the Mekong Delta, near the tip of South Vietnam, the Nui Ba Den Mountain rises straight up from the flat delta plain, the only mountain in that part of Vietnam. It's not far from the Cambodian border. The North Vietnamese and the VC controlled the bottom half of its slopes and U.S. and friendly forces controlled the top of the mountain. We had a PSP runway at the top, radar, and a radio relay site. I kept seven of my battalion soldiers there. They were on the mountain three months, then rotated.

The young W-1 flew by the mountain one day and the VC shot at his chopper from the mountain. He got ticked off. He started attacking those tunnels with the rocket pods attached to the sides of his chopper. He darn near ran out of fuel because he was torqued at the enemy for shooting at him. At the Army base on top, they had fuel. He descended, had them fuel his Cobra, went right back to the mountainside, and attacked the mountain again.

I heard about this happening through informal unit channels.

When his company commander heard about it, he came to see me.

"I'm gonna have to ground him," his Commanding Officer (CO) said.

"What tour is he on?" I said.

"This is his second tour, consecutive," the CO told me. "He hasn't been home. He's been here ever since he joined my company."

"I want to talk with him." I sent the kid's CO out to find him.

Later, I was out at the flight line, checking the maintenance on a couple of the choppers. We had a ready shack out there.

I saw this kid running out to get in his Cobra. He had a red scarf around him. The red scarf flowed three feet behind him. He crawled in the cockpit of the Cobra with the scarf thrown around him. The scarf was a problem because when you lower the canopy, the cockpit fills with air and part of the scarf flaps around inside. And that's what happened. Red scarf flapping away, he took off.

I guess it was the next day when I talked to his CO once more.

"What's this scarf baloney?" I said.

"It's his good luck charm," his CO said.

"Well, it hung in the canopy."

"The kid says he can't fail when he wears it." The CO cracked a slight grin.

I saw him land one day next to the Division CP carrying an unusual load, which caused quite a stir with the Division Chief of Staff.

"We have got to take prisoners," the Division Commander had said earlier. "We're not taking any prisoners."

I relayed the information to my three helicopter company commanders.

"Let your people know," I said, "if you have the opportunity, let's get a few prisoners."

That kid was incredible.

He was on a mission and three North Vietnamese ran to a hut. He machine-gunned all around the hooch and they came out, hands reaching for the sky.

He landed the Cobra, got out with his pistol in his hand.

"Drop those guns or I'm going to kill you," he said.

They pulled the guns off their shoulders and threw them on the ground. He tied them to the skids of the Cobra, two on one side and one on the other side. The rope was all that held them on the skids. He flew the prisoners back and landed on the Division helipad, at the Division Commander's CP.

He got out of the Cobra with the red scarf flowing behind him and strode in.

"Hey, I've got three prisoners out here," he told the G3.

The W-1 had been in Vietnam close to two years. He was near twenty years old because he was eighteen when he arrived in Vietnam.

"You've got what?" the G3 said.

"I have your three North Vietnamese prisoners out here." The kid pointed outside. "I was told you wanted them."

The G-2 ran out of the tent and the G-3 told me about it later.

"I've never seen such terrified prisoners," the G3 said. "The ropes holding them weren't tied tight, making them swing in the air underneath the chopper. The prisoners were hanging on for dear life."

"I tell you what," the G2 said, "I didn't have any trouble interrogating them. I told them I'd take them for another helicopter ride. They told me their life history."

"Bob," the Division Commander said, "I didn't mean to land one of those expensive Cobras out in the mud, have the pilot get out, and catch the prisoners with a pistol."

"Neither did I," I said, "but you don't understand the young man. He is absolutely fearless."

I told his CO I would talk to him.

"Son," I said, "you've done a splendid job here. As a matter of fact, I put you in for a Silver Star for capturing those three guys the way you did. But it's time you go home."

"I don't want to go home," he said. "This is where I'm supposed to be."

"Son, you've been here two years. You are one of the luckiest men alive. Your cat's got nine lives and you've already used ten of them."

"I want to stay here," he said. "Don't send me out of Vietnam. I love what I'm doing."

I dismissed him and talked to his company commander.

"He made a strong case," I told his CO. "He said, 'This is where I belong.'"

"Well, he is a heck of a pilot," his commander said.

"Tell him I'm going to give him three more months, then he's going home." I left to go back to my office.

But that didn't change his mental operation one bit. Haven't I already said it enough? The kid had no fear. The red scarf must have been one heck of a good-luck charm because he never got shot up, never got wounded, and he would do things that, if you thought about it, the average man would never do. He was a great chopper pilot.

The Cu Chi Night Light

We flew night missions around our perimeter at Cu Chi called 'Night Light.' The choppers went out about three miles every fifteen minutes and dropped a big aerial flare to illuminate the entire area. The operation was to harass the enemy on the ground.

The operations officer of the company battery was responsible for the Night Light mission. Four miles out of our firebase was an RVN Artillery Battalion with an American advisor. The ops officer contacted the Lieutenant who advised the RVN 105 howitzer batteries.

"Listen," the ops officer said. "Do you have any fire missions tonight."

"No," the Lieutenant said. "Nothing is scheduled. Tonight is an off night."

The ops officer sent out a Night Light chopper with a pilot, a co-pilot, and a crew chief. The crew chief shot the flares down.

The flares were mounted on the side of the helo. The crew chief would drop one of the big flares and it would illuminate. They flew in pitch-dark night. All of the sudden, something penetrated the belly of the chopper. It hit the co-pilot's feet, smashed his head through the top Plexiglas, and went through the rotor blades. The co-pilot had a bulletproof helmet on that kept him from serious brain damage. The Captain piloting the helo had lost his hydraulics and the cyclic stick was bent at a severe angle. Getting the daggone chopper on the ground in one piece was tough, but he made it work.

"Mayday, Mayday," the pilot radioed.

About the time he landed, one of my Hueys was out there. The crew brought the pilot, co-pilot, and crew chief to the base.

The co-pilot was the same man who had reported into my battalion with an overgrown mustache. He had grown it between the time he got out of flight school and the time he deployed. He had reported to Command Sergeant Major (CSM) Bennitt and me. I made an immediate decision.

"Sergeant Major," I said. "Take him out and shave him. I can't stand these mustaches."

The CSM took him out and dry-shaved him. He came in with blood spots all over his upper lip. Then I laid the law down to him. Now here he was, the co-pilot who had the shell impact under him, shoving his seat up, and blowing through the Plexiglas and the rotor blades without hitting anything else.

The shell, an artillery shell from the RVNs 105 howitzer, was a dud.

The RVN fired a high angle fire for a reason unknown to us and the dang shell slammed through the chopper.

The W-1 who was the co-pilot was spreading poison again. His buttocks were the color of my dark blue Levis. His entire backside was that color. The barrel of his .38 pistol was bent 180-degrees.

Believe it or not, he wasn't hurt except for the massive bruise on his tail-end. And his pistol barrel had been bent back on itself. My crew brought him back into the battalion headquarters and went back out.

"What happened to you?" I said.

I knew, because I had already been told. But I listened anyway.

"I'll give you a day off tomorrow," I said. "The day after tomorrow you're going to be flying your chopper again."

"You think I can, sir?" he said.

"I don't know whether you can or not," I said, "but you're going to."

I kept my word. After his day off, he was piloting on a sore tail-end.

"You can leave this country alive," I said later, "because you should be dead right now. Since you survived the artillery shell, you won't get anything closer than that."

The Warrant's company commander, the guy in charge of the artillery-damaged chopper, flew out to the RVN site and ate up that Lieutenant.

"What did you think you were doing?" the commander said. "You're firing artillery when you told me there wouldn't be any firing."

"Unbeknownst to me," the Lieutenant said, "The RVN

decided to shoot a high-angle fire."

"Were they shooting at our chopper?" the commander said.

"I don't know," the LT said. "I'll ask them."

They weren't.

But I was tickled with that Warrant. He spent the rest of his tour there and worked his way into a pilot's seat. He did a good job.

Awards

I received many combat awards in Vietnam. I won't go into all the details here as I've already mentioned a few of these actions before, but I will provide a short summary as these awards cover some of the most intense combat I encountered in Vietnam.

The Army awarded me two Distinguished Flying Crosses (DFC), the second award noted as an Oak Leaf Cluster.

The first DFC was for heroism while participating in aerial flight evidenced by voluntary actions above and beyond the call of duty while serving as the pilot of a command and control helicopter conducting an assault. During the assault, a well-fortified and concealed anti-aircraft gun emplacement shot down three U.S. helicopters. I drew fire away from the downed choppers while locating the exact positions of the helicopters and the anti-aircraft gun emplacement in a methodical fashion, risking my crew and myself in the process.

The second award of the DFC (the first Oak Leaf Cluster) was for valorous actions and exemplary leadership on 7 April 1970, which contributed immensely to the success of the mission during a daylong battle. In spite of intense enemy fire blocking friendly elements, I made continual low-level flights with the unit commander to pinpoint enemy positions. And I coordinated all aviation support for the ground maneuver elements, including air strikes, medical evacuations, and resupply missions.

For ground actions of meritorious achievement in connection with actions against a hostile force in Vietnam, I received two awards of the Bronze Star Medal, the first award on 6 October 1969 and the second award of the first Oak Leaf Cluster on 24 Jan 70.

I was awarded thirty-one Air Medals in my military career. Two of them were with the V-Device, which means for valor in combat action.

In addition to these awards in Vietnam, I received numerous other awards throughout my career. A list will be included at the end. For my service and actions in Liberia in 1980, the Army awarded me the Defense Superior Service Medal (DSSM). The European Command (EUCOM) Commander, a four-star General, presented the DSSM. But I will cover my time in Liberia and the Liberian crises later in the book.

I also received the Vietnam Service Medal with one Silver Service Star for the time I spent in Vietnam.

Though Hal Moore told me he would recommend me for a Silver Star, it never happened. But Hal was tied up fighting the enemy. He thanked me more than once for what he called "incredibly heroic actions in the face of danger without concern for personal safety." I'm just glad I could save his company from being overrun by the VC.

Vietnam II - LTC Gosney with 25th Aviation Battalion Sign (pg 121)

Vietnam II - LTC Gosney seated in Clipper 6. Specialist Anderson assesses battle damage for repair.

Vietnam II – LTC Bob Gosney (R) with brother CPT Gary Gosney (L) in front of Clipper 6 (pg 124)

Vietnam II – From L to R: LTC Gosney, CSM Bennitt, SP5 Anderson, CPT Nesbitt

Vietnam II – Cu Chi Night Light artillery damage to chopper (pg 147)

Vietnam II – LTC Bob Gosney, Commander, 25th Aviation Battalion in office (pg 121)

Liberia – 1978 American Embassy Party (pg 228). From L to R in civvies – Jimmy Don, Paramount Chief, COL Gosney, MG Koffa (pg 172), Liberian Army Commander

Liberia – 1978 American Embassy Party (pg 228) From L to R: Jean Gosney, Jimmy Don, MAJ Perkins, Liberian Army. Jimmy Don aide is in background.

Liberia – COL Bob Gosney in Paramount Tribal Chief Garb with Liberians

Liberia – Jean Gosney in Tribal Garb with Liberians

Liberia – Liberian Ministry of Defense Building (pg 191)

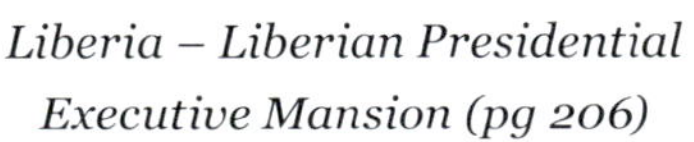

Liberia – Liberian Presidential Executive Mansion (pg 206)

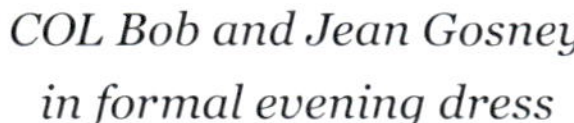

COL Bob and Jean Gosney in formal evening dress

CHAPTER 7

Liberia, Africa – An Unstable World

The Road to Liberia

"I've got good news and bad news," I said to my family when I returned to Hawaii from my second tour in Vietnam.

"What, Dad?" The kids gathered around.

"The good news is we can stay in Hawaii for another two weeks."

"What's the bad news?"

"We have to go to Washington, D.C. in the wintertime."

"Oh, Dad."

"That's where we've got to go."

We had to buy all the kids new winter clothes. In Hawaii all the kids had to wear were a T-shirt, short pants, and flip-flops. We bought them each new, warm clothes for the East Coast weather.

In Washington, D.C., I worked in the Pentagon as a Staff Officer, part of the Material Coordination Branch, Budget and Five Year Defense Plan (FYDP) Coordination Division, Office of the Assistant Chief of Staff for Force Development (OACSFOR), Department of the Army. I sat on a priorities board for Procurement of Equipment and Missiles, Army (PEMA) and was responsible for developing the ACSFOR position relating to the administration and allocation of PEMA funds amounting to four billion dollars.

Following the Pentagon assignment, the Army selected me to attend the National War College, a yearlong assignment to prepare future leaders for significant military and political-related positions.

After the War College, I completed assignments at Fort Hood, Texas until 1978. My memorable assignment at Fort Hood was as the Chief, Operations Division, G-3/DPT (Director of Plans and Training). I was responsible for supporting testing of units, Reserve and National Guard Annual Training, the Corps/Fort Hood Emergency Operations Center—a 24/7 operation, acquisition of off-post training land for maneuvers and flying, and monitoring the readiness status of a hundred and fifteen units. I performed that duty in the most visible and demanding Lieutenant Colonel staff position at Fort Hood in an outstanding manner according to my bosses.

One of my most notable assignments was to Liberia as the

Chief of the Military Mission; however, Liberia wasn't the initial projected assignment for me. In Africa, the nations and their politics changed at a rapid pace. The assignment to Liberia was not a simple admin process, but a response to fulfill the United States mission overseas.

The Assignment Process

My full-colonel assignments officer had slotted me to fill a Brigadier General position in Ethiopia. I was excited and already on orders.

A friend of mine, BG Bill Wing, currently filled the position. I drove to the division headquarters and used their secure phone to call him.

"Bill," I said, "this is Bob Gosney. I've been assigned to take your job."

"Well, Bob," General Wing said, "I have a hard time telling you this, but Mengistu, the Chairman of the ruling Board, the Derg, is kicking us out of the country. My staff and I have already shipped our stuff out of here. We'll be leaving in forty-eight hours and we'll close the Embassy."

I called my assignments people on another phone, an unsecure line.

"Do you still want me to go to Ethiopia?"

"Yes," my assignments officer said. "We need you to go. It's a critical post."

"I don't even have an office there," I said.

"They'll have one for you."

"Have you talked to General Wing lately?" I said.

"No, I haven't."

"Do you know what's going on in Ethiopia?" I said.

"No," he said, "I don't."

"Well, you'd better check."

"Why?"

"This isn't a secure line," I said, "but you call and find out, because I'm not going there."

"You're not?"

"No. I'm not going over there. But if you hurry, you can talk with Bill Wing."

The phone line was unsecure. I couldn't tell him all the information I had heard. He called back in fifteen minutes.

"I'm glad you let us know about the situation," he said. "General Wing told us he had sent a letter by diplomatic pouch a week ago telling us the news. I guess we didn't read it."

"Don't worry," I said. "I wouldn't have gone. They'd all be gone anyway."

"We've got find you a different assignment," he said. "Hey, I've got a position in Liberia."

"Liberia?" I said. "That doesn't sound good."

"Oh, no, it's a good post."

"I guess it will do," I said.

The assignments office cut new orders and rescinded the ones for Ethiopia.

That's how I wound up in Liberia.

Liberia—A World of Tension

Liberia was the most fascinating military assignment of my life.

The person whose position I was taking had already departed. He was persona non-grata. The government of Liberia didn't want him there. After a few weeks there I could see why. He might have been gruff, not establishing any rapport with the military or with the civil authorities because they didn't have anything good to say about him.

Liberia was a whole different world. Beyond the difference of culture, I had to deal with oppression of the poor, graft, and other chronic problems. Even simple transactions could be made complex because of the criminal activity of high-ranking officials.

And I found out real quick all blacks weren't alike to the Liberians, though that was their own skin color. The two major groups were the Tribals, who were the native Africans, and the Americo Liberians, who were descendents of American slaves. The Americo Liberians were a minority in power, and most had lighter skin color than the Tribals. The current Liberian President, William R. Tolbert, Jr, was a member of The True Whig Party, the minority Americo Liberians who were in power.

When it came to the whites or Caucasians, we were an even

tinier minority. Because of the cultural, political, and social inequities, Liberia was the most exasperating and exciting tour in my military career.

Military Mission to Liberia

My title in Liberia was Chief of the Military Mission or Milmish for short. I worked directly under the Ambassador.

I was in charge of seven officers on the Milmish staff and their families. We all lived in compounds, which were collections of homes bordered by concrete or barbed wire fences. Jean and I lived in a house about ten times bigger than the size our current home. It was a beautiful set of quarters.

One of my jobs was going to foreign embassy functions. We'd get invited to all of them. I didn't like the parties because my hearing was going bad and I didn't have any hearing aids. I could read lips or give a polite smile.

The Chinese conversations were blunt to the extreme. They would ask for classified information.

"How many M-16's do you have?" they asked. Or, "What kind of weapons do the Liberians have?"

"I can't tell you now," I told them.

Sometimes other Europeans would talk with me and my response was a grin or a nod. Though I didn't enjoy that part of my job, it was a requirement.

My real job was to train and equip the Liberian Army, Navy and Air Force. They had a Ranger Battalion (BN), which I formed when I got there. For U.S. troops, I had a Naval officer, an Air Force officer, an Army officer, and a Special Forces A-team that rotated on a periodic basis. A Special Forces A-Team consists of eleven top-notch soldiers, commanded by an officer. There's a Non-commissioned Officer in Charge (NCOIC), a Medic for medical issues not requiring a doctor, a linguist who could speak the language, several soldiers who handle firepower requirements, a communications soldier, and an armorer. They were excellent at their specialty and cross-specialties—well trained.

The Liberian Air Force didn't consist of any war machines—they had several push-pull aircraft and L-19s, which are Cessnas.

When I first arrived, there were no difficulties. We didn't have any uprising or major issues to deal with.

The Liberian soldiers were good. But they had a memory half-life of about two weeks. I had a U.S. Special Forces training unit working with the Liberians. During compass training, the SF soldiers would tell them to walk north using the compass. They'd teach them various things about the compass.

Two weeks later, the instructors would ask the Liberian students a question.

"What do you know about this compass?" the SF soldier would say.

And the Liberian soldiers wouldn't know what to do.

Settling Into Life In Liberia

Liberia had two of the largest rubber plantations in the world. When our airplane landed in Liberia, our first smell was the stench coming from the rubber. Jean, my wife, and Amy, our youngest daughter, accompanied me to Liberia, experiencing that initial blow to the senses. The horrible stink from the rubber plants did not make a good impression on us. However, in no time you get used to the smell.

Our home in Monrovia, the capital of Liberia, covered about ten thousand square feet. The downstairs had bars on the windows and big sliding doors to the yard. We had a bathroom outside. Upstairs the windows had no bars and a big balcony overlooked the ocean, which was near our home. You slept at night with the lapping of the waves lulling you asleep. "Heck, this is a better deal than Ethiopia," I thought. Jean enjoyed living there. Personal items to make our house a home—pictures, knick-knacks, and other personal touches—we brought to Liberia, because the government had furnished the house assigned to us. We stored most of our household goods in the States.

Our houseboy was SGT Alfred. He idolized me.

"If Colonel Gosney told me to jump off a building, I'd do it," Alfred would say.

I made Alfred a deal about painting the house.

The building was made of concrete tile about twelve to fourteen inches thick.

"I want to you to paint the whole outside of our home," I said to Alfred.

"What do you want me to pay the workers?" Alfred said.

"I'll give the money to you and you pay them." I handed him five dollars, which was a lot of money to them.

Health-wise, we didn't encounter any major problems. Though Liberia had a huge Ebola virus outbreak in 2013, one of the largest in Africa, we didn't have the Ebola virus problem when I lived there.

Liberia contained the largest radio intercept in Africa. The footprint covered a thousand acres and was run by a U.S. agency. The facility, called the ELWA (Eternal Love Winning Africa), was out in the bush.

The cops in Liberia didn't have police cars. They'd commandeer a little taxi and away they would go. One day I was at the beach with the dogs. A yellow taxi arrived at our house and Amy, my youngest daughter, was in it with her Dutch boyfriend. A policeman was in the car, too. Amy wasn't concerned, but the Dutch boy was scared.

"I give these two a ride and they owe me money," the policeman said when I arrived.

"Get the heck out of here," I said. "Now."

He backed into the cab and sped away.

"He brought us home because he wanted twenty dollars," Amy told me later.

The police were corrupt. That policeman had commandeered a cab, left the driver standing, and drove the cab to our house.

Nothing worked right in Liberia. Nothing. We had generators because of black outs. The power stopped without warning and the generators purred to life. We had to have constant backup power to keep the lights working and keep our deep freezers and refrigerators operating.

We ordered our groceries from Charleston, South Carolina. The military flew the food to us in C-141s for all the American military staff. We sent a telex with a grocery list and received a delivery via C-141s. The planes delivered food once a month. We got plenty of groceries.

I laughed a lot. I'd say the Liberian national bird was a little white car left on the side of the road. The Liberians would say it was spoiled. The problem might be a mechanical issue, but they didn't get them fixed. And, of course, the jungle growth swallowed them up.

Discovering and Challenging the Culture

Even though 1978 was a quiet year in Liberia, I could sense the underlying inequities of the Liberian government when I arrived there. For instance, the United States sent shiploads of rice to Liberia, not for sale, but as a free gift for the Liberians. A Liberian with three kids needed rice because it was their staple.

I visited the port one day and two Liberian LTs and a CPT had a stand selling U.S. rice to soldiers. I did like Jesus did—turned the table over and ran them away. I phoned two of my U.S. soldiers.

"Make sure the Liberians give this rice away free," I said.

I found out the money from rice sales went to the Chief of Staff, who kept a small part and gave the majority to the President.

Corruption was rampant and proved to be a major thread in the unraveling of the country's leadership. You had to watch the Liberians like a hawk. 'Dash' was a type of bribe you paid to get anything done. I refused to pay and I still got what I wanted.

Advising the Liberian President

After I arrived in Liberia, I met with President Tolbert four days a week in his office. When I talked with him, both of his legs bounced up and down due to nerves.

"Mr. President." I leaned forward in my chair. "If you don't change the way you treat these soldiers, you're going to have a coup. You need to have a Post Exchange (PX) where they can buy things cheap. You take twenty-five dollars from them and they get paid a total of forty dollars a month. You know, grief overwhelms the soul. The soldiers don't have enough to live on."

The tension was high on that visit. And sure enough, he got killed in the 1980 coup. Instead of taking most of the money, he could've taken half of it and improved the outlook for all his soldiers.

The Liberian Batmobile

As an advisor to the Liberian President, I spent a great deal of time close to him. Every time he left on a trip, he took me with him.

One of the President's vehicles was the Batmobile. The car had big iron running boards on the sides with a bar along the roofline of the car for soldiers to hold on. The lead man on both running boards had a long, thick iron pole. If a car didn't move aside fast enough or far enough, the soldier thrust the pole out, crashing it into the offender's windshield and knocking it out.

We were coming back from a trip, headed toward Monrovia and I saw my wife driving our car. With a sudden jerk, Jean steered the car off the side of the road. I was sitting on the right side of the Batmobile. I waved as we went by. She didn't wave back. She was afraid because she had seen those soldiers break windows and cause wrecks. I grinned as she sped away.

The President used my presence during these rides as a show of his power. The Liberians had an intense concern with the appearance of might. The President asked me all types of questions, which I answered without problem. He spoke pretty good English.

President Tolbert's Wives

President Tolbert had fourteen wives with one-hundred-and-fifty-two kids. He had one Christian wife according to the way Liberians described his relationships. The Christian wife was the one who lived with him. The rest of the wives were called country wives. They didn't live with him, but he put a roof over their heads. There's no telling how many he had relations with. He asked me what I thought about that many wives.

"I've got one wife," I said. "She's hard enough to handle. I don't want fourteen more." I chuckled.

"Is she a big woman?" He looked at me.

"No," I said. "You've met her. She's a little woman."

"You don't know what you're missing." He shook his head.

The Liberian President's State of the Union Address

Each year the Liberian President made a State of the Union address.

President Tolbert had experience speaking. He was a pastor in the Baptist church, which didn't mean a heck of a lot, but was a position of great authority. The Baptist church he attended was big. His pastoral position made him a preacher as well.

Tolbert's State of the Union address was a spectacular affair with hundreds of dignitaries. He had all of the Liberian government officials attend. Tolbert always had me sit on the front row next to the Chief of Staff and to the Army Commander.

His State of the Union address went for six hours, nonstop. That was the most miserable six hours of my life. He talked, and talked, and talked. Of course I had to look like I was interested in what he said, but I was in absolute distress sitting there in the open air. It was the Harmattan season where it doesn't rain, but it was hot. We sat outside because they didn't have a building big enough to hold all the people.

He started with a thirty-minute long prayer, then launched into his State of the Union speech. In his address, he said the Union was in wonderful condition. Unbelievable. That wasn't true at all. His speech sounded like a certain American President I know, but I'll let you guess which one. Tolbert went on forever. I glanced over at the Chief of Staff and his head was nodding. I jammed my elbow against him to wake him. I thought, "If I'm going to sit here through this miserable thing, you are going to stay awake."

I went through two long speeches of his. I didn't enjoy either of them. If he had talked for an hour-and-a-half, it would have been bearable, but six hours was ridiculous. I checked my watch when I sat. The event started at one o'clock and ended about seven.

After the State of the Union address, everyone grazed through a layout of Liberian food. I liked their food. They served rice cooked thick, a soup to put on the chicken and on the rice, hot peppers, and many other dishes. But after sitting six hours, I had lost my appetite. Despite the lack of hunger pains, I ate a bowl of rice.

"Well, Colonel?" President Tolbert approached me. "What did you think of my speech?"

"Mr. President," I said. "You were magnificent. Wonderful. But did you ever try to shorten your speech a bit?"

"I did." He gave me a smile.

"I hope you did." I said. "If you hadn't, it would have gone a lot longer."

"Well," he said, "I realized I had to cut it back."

Liberian President's Lack of Caring for Soldiers

President Tolbert often asked me about his soldiers.

"You are abusing those soldiers," I said. "They're getting mighty restless."

"What do you mean?" the President said.

"You've educated some of them," I said, "The educated ones are telling the uneducated ones how bad their situation is because you're taking too much of their pay. They need a place to buy inexpensive items. Another thing is the donated shipload of PL-13 rice from the United States. That rice is to be distributed to the military people, free. The United States government sets the program. It's free for military families."

The Liberians had 12,000 active duty military in Monrovia. They had 30,000 reservists who lived scattered around in the villages. Each soldier had weapons and ammunition.

"You are aware, aren't you," I said, "that your active duty and reserves are a potent force. Don't keep them edgy and feeling like they're being abused."

Restless Liberian Soldiers

The soldiers were getting restless. They talked to me.

One day I had to go to a camp and speak to two battalions of Liberian infantry. They were on pins and needles. I had already told the president of about this tension.

"I'm going to talk to your President and see what we can do to improve your plight," I said to the soldiers.

I made sure they understood I was serious.

"You have a legitimate complaint." I paused to gather their attention. "But good soldiers don't start uprisings and create havoc. Put your faith in me. I'll represent you."

The soldiers all clapped, whooped, and hollered.

That's one time I told President Tolbert about the issues. We were on a drive and I told him what I told the two battalions of infantry soldiers.

"You need to support me on this," I said. "We have to get them under control."

In the car he bounced in the seat, his feet tapping the floor. He was that nervous. But greed is an overwhelming thing. And Tolbert had greed galore.

"How much is enough money for you?" I said.

"I have a lot of expenses." His feet kept tapping.

He didn't have a lot of expenses because of all the building he had done. He was the President of a British building company, owning 51% of their stock. His position in the company gave him 51% of the profits, which was a pile of money.

I knew how rich he was. Heck, his palace in the country had gold chandeliers, gold faucets, and the other gold touches in the plush bathrooms. He had his house furnished with a variety of opulent furniture and finishes. But his palace was dirty looking compared to the guys living in a thatched hut on a dirt floor.

A Liberian Post Exchange

Days or weeks after these conversations President Tolbert called me to his office.

"What I plan to do," Tolbert said, "is to build a PX (Post Exchange) one day and I'm going to stock it with things the soldiers can buy below cost."

"Prices have to be below cost without any kickback."

"Agreed," he said. "I concur."

"I don't see a need to build a new one." I sat a little straighter. "Out at camp Schieffelin, there's a new building sitting vacant. The structure would be an ideal place to put a PX, because that's where 90% of the soldiers are." I smiled. "If you're going to build a new PX it will take three or four years, because in this country you have to order everything. The building materials will be shipped in from other countries, which takes forever.

But he never took action. The PX didn't get built.

U.S. Takes Command of the Liberian Army

The country ran smooth until one day, April 14, 1979, when everything went to hell in a hand basket.

There are five tribes in Liberia. When leading or managing a nation consisting of tribes, a concern is the people's allegiance is to the tribe and not the country. In other words, Liberia's five tribes had no allegiance to Liberia.

The tribes thought they could come to Monrovia, using the front of the quasi-Marxist Progressive Alliance of Liberia, to stage a peaceful march on the capital. But in reality, the tribal people's intent was to rob, plunder, rip off whatever they could take, and return up-country with it. What started as a peaceful demonstration soon got out of control—total chaos. The tribes-people broke into Liberian homes and a great number of Lebanese merchant's homes, stealing everything they could get their hands on. Most of the rioters had M1 rifles. Thirty thousand Reserve soldiers who had lived up-country carried most of the rifles.

As the riot became a total mess, President Tolbert held a meeting with Jerry Richardson and me. Jerry was the CIA Chief in Liberia. We met in a building surrounded by guards. Tolbert's staff was there, including the Minister of Defense, the Minister of Foreign Affairs, the Chief of Staff—Henry Koboy Johnson, the Army Commander— MG Stephen Jaitoh Koffa, and many others. The civilians were scared to death. Their legs shook.

"Can't your Army put this riot down?" I said.

Tolbert realized the riot was out of control. There were several thousands of these renegades with their M1 rifles coming into Monrovia and 'shopping'—all free.

"What can you do about it?" Tolbert looked straight at me.

"Mr. President," I said, "there's not a lot I can do. I can organize the Liberian Army, but your generals and colonels have to command them. They must get them to do what they're supposed to do."

"I want you to take command of the Army," Tolbert said.

"It's against the law in our country." I shook my head.

"I'm giving it to you and I want you to command the Liberian Army." President Tolbert's legs bounced. "Convey that message."

"I'll go to the Embassy." I nodded toward my CIA compatriot. "Jerry and I will will see what can be worked out."

We drove to the American Embassy and I talked to Ambassador Julius Walker.

"Julius," I said, "the President wants me to take command of his Liberian Army."

"You can do it, can't you?" Julius said.

"I sure as heck could. But it's against the law to command a foreign army."

Julius phoned our Defense Department.

"This is Ambassador Walker," he said. "I need to talk to the Secretary of Defense."

After a while, Julius was talking to him.

"We have the Liberian President here who has told Bob to assume command of the Liberian Army."

"What?" said the other person. Even I could hear the response. He was startled.

"Yes," Julius continued, "and it's important Colonel Gosney assumes command."

"Let me talk to the Secretary of State."

Julius waited as the Defense Secretary spoke to the U.S. Secretary of State. Jimmy Carter was the President of the United States at the time. I guess he talked to the Secretary of State. I sat there with Julius, waiting it out. It took about thirty minutes.

"He has the approval to do it," said the Secretary of State.

"Approval? How high?" Julius said.

"The President."

"OK." Julius turned to me. "Go tell them you'll take command of their Army."

Jerry and I drove to the guarded building housing Tolbert

and his staff, strode into his office, and waited to get his attention.

"I'm going to take control of the Army, President Tolbert," I said. "But I want Koboy Johnson out of the way. And get BG Sheffield out of the way." I swung my gaze to my left. "Major General Koffa, you can take a few weeks off."

I got rid of the riff-raff leaders who were stealing money. I sent my Special Forces team to Camp Schieffelin, a big military camp between Monrovia and the Roberts International Airport.

"I'll be down there," I told my U.S. Special Forces leader, "Get twenty two-and-a-half-ton trucks. Have the Liberian soldiers dressed in combat uniforms." I issued exact directions to organize the soldiers to have the best effect in this riot situation. "Tell the officers to fill each truck with the soldiers. Have them fix bayonets to the M1s, put steel pots on their heads, and sit at attention. No ammunition on those trucks. None."

When I got to Camp Schieffelin, the Special Forces had started the process and had it pretty well organized. I had one U.S. Special Forces soldier ride in each truck in the cab with the driver.

I had twenty trucks with soldiers sitting at attention, bayonets pointed in the air, and with their steel pot helmets on. I went around, checking each truck.

"I told you to sit at attention," I said to a slouching soldier. "You sit at attention." The soldiers looked stiff as boards.

"Okay. Each of you take a quadrant." I pointed to the leader of each five-truck section. "South, North, East, and West. Roll around downtown Monrovia."

Within three hours the riot was over. The rioters weren't aware the soldiers didn't have any bullets. But the way they sat, each soldier appeared to have ammo.

The whole idea was to create a show of force. Most of the rioters and looters headed up-country, though small numbers of renegades stayed in town. We had a special way of dealing with them.

Restoring Order After the Rice Riots

After the riot was under control, people still flowed in from up-country to plunder. The criminals brought their weapons with them. The Lebanese owned most of the businesses, but the President had 51% of the Lebanese business as well as the British business I mentioned earlier.

To stop the stealing, most of which occurred after hours, I established a curfew. The requirement went something like this: Anybody who came into Monrovia or the immediate area surrounding the city had a curfew. Everyone had to be off the streets at 8:30 P.M., except for law enforcement or Army personnel. Anyone found on the streets after the curfew would be subject to punishment.

"Don't shoot them," I told my Army soldiers policing the streets. "Whip them. Whip violators good when you've caught the rascals." The whips were made from certain parts of the bulls. They were pretty long and boy, they made a heck of a whip.

One night I was at the El Meson Hotel on Carey Street, which was the single Monrovian hotel where you could live without rodent problems. Monrovia had a couple of other rat-infested hotels.

I was on the restaurant balcony of the hotel. The structure was three-stories and we were on the second floor. Jean and I had friends who owned the place, Manuela, a Spanish woman, and her Lebanese husband, Toney. They did an excellent job running the establishment and had a fine restaurant. I had stopped by to visit them and we sat overlooking the streets.

Unbeknownst to me a stringer was at the restaurant. I thought he was a friend of Toney, who thought he was a friend of mine. A stringer is a newspaper or magazine guy who goes to newsworthy places to make money writing and selling stories.

On the street below us three Monrovian soldiers had stopped a man violating the curfew. They whipped the thunder out of him with their bullwhips.

"Brutality," the stringer said. "Horrible, horrible, horrible..."

"What's so bad about it?" I said.

"They're whipping that man." He pointed at the action below.

"Do you want me to tell the soldiers to shoot him?" I raised an eyebrow. "They're all carrying their pistols. They could shoot him if they wanted to. Would it be better if I told them, 'Shoot the guy, don't whip him?'"

"Are you serious?"

"Yes." I narrowed my eyes. "That's why they're whipping him, because I told them to punish violators without shooting them."

I explained the enforcers would whip the individuals and haul them off to a jail for the night. He couldn't believe it.

"Who in the world are you?" I said, testing him.

"I'm a stringer."

"Then shut-up. The guy on the street isn't going to be killed, he's going to be whipped."

And my Lebanese buddy, Toney, was at his side in three swift strides.

"Get off of my balcony."

He grabbed the stringer by the shoulder and led him down the stairs.

"You better watch out, because you're violating the curfew. Now move fast and find your home where you're living."

He pushed him out on the street.

I saw the stringer racing from building to building, hiding in the shadows. I looked for the man's story to come out in the papers, but I didn't see it.

If I hadn't put out the edict that soldiers couldn't shoot any people who were violating the curfew, who knows what kind of mess we would have started. My rules were for their benefit. But I told them absolutely no shooting. As long as they could take action by bullwhipping the culprits, they had something to do. The soldiers listened. When I told them something, they listened to me and didn't violate my orders.

I kept command of the Liberian Army for three months in order to get it organized for other things.

CHAPTER 8

LIBERIA, AFRICA – FIGHTING CORRUPTION

No matter which direction I turned, I found disorder and dishonest people at all ranks. Though the Liberian nation had suffered a significant emotional event when the 1979 Rice Riots occurred, the international companies working in Liberia didn't change for the better, nor did the decisions of high-ranking Liberian officials and diplomats. I applied fixes where I could and even explained the consequences of corruption to the President, Minister of Defense, and the Chief of Staff of the Army. However, this did little to curb the behind-the-scenes actions of greedy individuals. In spite of this, a few incidents still brought a smile to my face.

Voice of America Party and the Phone Line Connection

President Tolbert tried to be a pleasant man, but I knew he was the greediest guy in the world. He got a piece of the action from every angle, even the telephone system.

The British ran the telephone system, which seldom worked. My phone at home was connected, but I couldn't talk to anybody at work. We had little Motorola radios for communication with repeater antennas around the country. I could talk on a radio to my guys a hundred miles away using the repeater stations, even somebody in Bonza, the deep jungle. But not having telephonic connection to my office from home was an issue for me and evidence of a principle of poor service from the British phone company.

Sometime before 1980, I decided to fix the problem through the Liberian signal corps since I controlled the training of the Liberian Army.

"Run a phone line into my home that will work," I said to them.

The Signal Corps came out to my house, ran a phone line through the British phone line system, and connected it to my office phone. My phone worked great.

After this, Jean and I were invited out to ELWA, a compound where Voice of America (VOA) had a business operation, for a party. VOA employed a lot of civilians and their wives.

While we were there, I met a British guy.

"What do you do here?" I asked.

"I run the telephone company here in Liberia," he said.

"You do?" I tilted my head. "Why in the world doesn't the phone system work?"

"I'm working on it." He nodded with British self-assuredness.

"My phone is working now. No thanks to you."

"Well, old chap, how did you do that?"

"I had the Liberian Signal Corps lay a new line from the switchbox all the way to my house." I locked eyes with him. "And my phone works."

"You can't do that," he said.

"What do you mean I can't?"

"I'll send people out to cut your wire down." He wasn't smiling now.

"Go right ahead." I made a dismissive gesture with my hand. "I'll shoot their tail ends off the pole if they come out to cut my wire."

"Are you sure?" he said. "Are you serious?"

"I'm damn serious." My jaw was set firm. "If they climb my pole and mess with my phone, they'll have it coming to them."

That shut him up real quick.

"Why don't you come out yourself?" I asked. "If you climb the pole and try to cut one of my wires, I'll shoot your tail, too."

They never did come out to cut the wires.

I'll bet I had the only phone working in Liberia because the Liberian Signal Corps put new lines in for me. When I told the man I would shoot his employees, he was aghast when he realized I was dead serious.

From that point on we had pretty good phone service at our house. However, I don't know how other people's phone service worked.

Fixing the Liberian Soldier's Pay Problem

I told President Tolbert I was going to send my officers to all the military pay stations. The Liberians paid their soldiers like Americans paid a battalion back when it was done with pay officers visiting the units. The Liberian officials arranged the soldiers in a line. Using a roster listing the soldiers by name, they called them forward to pay them fifty dollars a month.

The soldiers got paid. However, the Liberian Army Lieutenants, who were acting as paymasters, weren't paying the soldiers all the money owed them.

I sent my guys to each pay station. They came back and broke the bad news.

"It was awful," one of them said. "Out of fifty dollars, each soldier was lucky to get fifteen dollars because the Lieutenants would say 'this much is for Liberian taxes, this much is for county taxes,' and the list went on and on."

I went to the President and told him what I had found.

"What can we do about it?" he said.

"I'll stop it," I said, "because my officers are going to be the pay officers for all the soldiers."

Tolbert grimaced. He didn't like that one bit, but said nothing because he was getting the biggest kickback from the bilking of the soldiers.

After we talked, procedures changed. The Liberian Lieutenant sat with my guy right beside him. When it came time to pay the soldier, the Lieutenant paid him fifty dollars in cash. Soldiers received their full pay. A Master Sergeant got about sixty. Sergeants got a little more than fifty. Privates got fifty. The pay scale was graduated by the increase with rank, but not by a heck of a lot.

The Liberian Chief of Staff was incensed because he got a cut of the soldier's pay. The Chief of Staff's office was next to mine in the Ministry of Defense building. I stopped by to talk.

"We stopped the Liberian officers from ripping off the soldiers pay," I said. "They're getting all of their pay."

"All of it?" he said.

"All of it." I nodded.

He didn't say a daggone word. He knew I was aware he got an under-the-table, brand-new Chrysler every year. I stopped that action as well. A man from an Egyptian company gave him a car as a bribe for the uniform contract every year. If he got the contract, the Chief of Staff got the car.

What surprised me was my actions were like trying to put out a grass fire with a fire extinguisher. All kinds of little areas had illegal activities. As I cut out the graft, I was amazed at the Army's increased efficiency, because the soldiers were happy. They loved getting their entire paycheck. And a happy soldier is a good soldier.

Cutting Graft from the Uniform Contract

I fixed the uniform contract problem, too. I created a program where the companies who wanted a uniform contract submitted a proposal with an estimate of how much the contract would cost. I told those coming in with bids what to do.

"I don't want an estimate," I said to them. "Tell me the charge per uniform."

Each bidder got the same information. Five people representing various companies submitted bids and proposals.

To make the best decision, I had a group of top-notch, efficient Liberian officers join one of my Lieutenant Colonels and me on the governing board. The low bidder got the contract, which saved costs and kept greed from ruining contract operations.

I found out first hand how the corruption had occurred.

The Egyptian Ambassador invited Jean and I, and the Station Chief and his wife to his home for a dinner reception. I met a distinguished looking Egyptian whom I thought was an officer in the Egyptian Embassy. He was a handsome, tall gentleman named Naha.

Naha made it a point to converse with me often during the reception. I thought, "This is odd. He's paying too much attention to me and not enough to the Liberians who had attended as his guests."

The next day in my office, my secretary, the Ambassador's daughter, came in.

"There is a man from the Egyptian Embassy here to see you," she said.

"What's his name?"

"Naha." She waited for a response.

"Oh." I remembered him from the party. "Send him in."

I learned he wasn't from the Egyptian Embassy—he represented a uniform manufacturing company in Egypt.

"I want the contract," he said flat out. "I've gotten it every year for a long time."

"Times change," I said. "The process is different. You submit your proposal with a bid. Your submission will be put with the others. We'll select the cheapest bid."

"How does five hundred thousand dollars sound to you?" His eyebrows lifted with a conspirator's intent.

"Is that your bid?"

"No." He shook his head. "I'll give you five hundred thousand dollars for the contract."

"Are you trying to bribe me?" I acted like I was peeved. But it didn't take much acting. I was.

"Oh, no." He paused. "I do this all over Africa."

"Well, you don't do it here." I said.

"Six hundred thousand dollars?"

"No. You don't do it here." I made it clear—no bribes.

I went home that night and told Jean.

"You turned it down?" She laughed, stopped herself, and laughed again.

Because I told her Naha had said, "This will only be me and you."

"Naha," I had responded, "I've been around here, to Fort Hood and Dallas, Texas, too. When two people know it's a secret, it's no longer a secret. No, you will not pay your way into this contract." I was referring to the Chrysler he used to bribe to the Liberian Chief of Staff, which was cheaper than the money he offered me.

His face had the saddest expression, crestfallen, because he couldn't bribe me.

Before the deadline closed, we had the Egyptian's bid along with all the other company's bids. The board sat, reviewed the bids, proposals, and final amounts. Another company got the contract. The competing company built a low bid and the proposal was the same. In the end it would be a quality uniform for a lower price.

The contract proposal process I instituted was a revolutionary concept to contractors working with Liberia because it cut out the availability of kickbacks. The Liberian officials' greed still remained. The Chief of Staff's new Chryslers quit coming. He had to drive the last year's Chrysler for a long time.

"I'll tell you, Colonel," said a political officer at the American Embassy, "you're going to tick off a lot of people."

"I hope I do," I said. "Because they're not going to rip off the Liberian Army. They're not going to be greedy with those contracts."

I was there for two and a half years and every year we did the same thing. It's funny, though. The President never said a word about the contracting process changes. I would've come back at him real quick, but he didn't comment on it. The old Chief of Staff didn't say anything either after I told him what I had done.

In a country like Liberia, the President got the lion share of the ill-gotten gains and then it dwindled from there. The President received at least a 75% cut of all the graft. And other lesser dignitaries got their cut, sometimes in expensive goods, like the Chief of Staff's new Chryslers. But I put a stop to it. Not on my watch.

"This is the way it is," I said. "And this is the way it's going to be."

The African Unity Conference—July 1979

The African Unity Conference, held July 17-20, 1979, in Monrovia was for all the African nations in the Organisation of African Unity (OAU). Haile Selassie I established the OAU in 1963 in Addis Ababa, Ethiopia. The meeting places varied each year among the nations. In 1979, Liberia was center-stage for the first time.

President Tolbert had his British company build fifty-one houses for the attending dignitaries from each African nation and a convention center. Each house had a brand new Mercedes Benz for the dignitaries and lots of other ludicrous benefits. Of course, Tolbert was a 51 % owner in the company and reaped a great deal of profit. I wasn't involved in this activity and didn't want to be involved. The Africans staged the entire affair.

The Liberian Army set up a parade and other activities for the Conference.

As I watched this transpire, I questioned what was happening. Where in the world was the Liberian government getting the money to do this? I never found an answer.

The construction turned out modern cottages. Each cottage had a brand new Mercedes Benz. That meant the government had purchased fifty-one Mercedes, which other people were driving.

The payment of the costs for the entire conference made me curious. The country is composed of about 99% poor. The poor citizens are peons because the Liberian government keeps them under the feet of the state.

After the conference, a lot of the soldiers who were part of the later coup in 1980 moved in the cottages. One of soldiers wanted me to come out and have food with them. I had lunch with about seven of those coup-boys, as I called them.

"You need to take care of these houses," I told Samuel Doe when he became President of Liberia.

"Oh, we will, Chiefo," the President said, "we will."

As it turns out, Doe killed all but one of those coup-boys. He was afraid of a counter-coup. But they were mere helpers, poor African soldiers who supported Doe.

Liberian African Unity Conference Motorcycle Escort

In preparation for the OAU conference, the scene in Monrovia was like the old Keystone cops movies or a slap-stick comedy. The Germans sent a team to train the Liberians to drive motorcycles. They were to train the Liberians how to do a motorcycle escort to create a credible escort for the attending dignitaries during the Conference.

The Keystone Cops might be too nice of a description. I was

driving to the Embassy on a sunny day and a motorcycle roared past me. The man driving the motorcycle must have recognized me because I was in a sedan. He tried to look important. He kept his eyes focused on me and waved as he shot past. Then he faced forward and nailed a Volkswagen.

I laughed.

He somersaulted through the air.

The Germans quit. The Liberians wrecked eight or nine motorcycles. The Germans are perfectionists. They shut down the training. The Liberians had to use cars to escort their dignitaries. You couldn't teach them to drive a motorcycle even if they took another year of training.

Arabs Hijack U.S. Communications

On top of the Edwin J. Roye building on Ashmua Street, known as the EJ Roye building for short, I had one of my repeaters for our communications network. The repeater radio was housed by a shack, which also included the relay and the rest of the comms gear.

I began to notice on Wednesdays, about noontime, our radios would go off the air. Jerry Richardson, the CIA agent and a great friend, was with his wife, Virginia, at the El Meson restaurant. Jean and I were having dinner with them. I had my mobile phone open, and all of a sudden, I noticed it had quit.

"Jerry," I said, "I'm going to the EJ Roye building, get on top of the dang thing, and see why our repeater is not operational."

"I'll come along." He stood. "Let's go."

We got the ladies ready. Then we drove to the EJ Roye building and parked in front. He and I hustled inside. The elevator had gone to the top floor and hadn't returned. I pushed the button several times, but it wasn't running in either direction. The level indicator showed the top floor.

"The elevator's stuck," I said.

"Let's hoof it." Jerry pointed at the stairs.

We sprinted up those stairs, the long way to the top. When we reached the roof, I checked out the repeater building. The door was open. I hurried over and Jerry followed.

I stepped inside and saw three Arab-looking men sitting with a radio hooked to our antenna, leaving our system disconnected. One man was used their radio.

"Who are you?" I said.

"Are you speaking Espanola?" He tilted his head to see me.

Looking at those men you could tell they weren't Mexican.

"You're not Hispanic." I roared back. "You're a daggone Arab."

He started to speak, but I almost beat him to the punch.

"Come one, Jerry," I motioned him forward. "Let's put these guys under arrest."

They bolted, running like they were on fire. They took the stairs, arms and legs churning.

Jerry and I used the elevator. It was stuck because the door wouldn't close. The trio had stuck a stick between the door and the elevator frame. We removed the stick, got in the elevator, and almost beat them to the lobby.

As the elevator doors parted, the fleeing men dashed across the lobby and out the front door. Jerry and I rumbled out behind them like another Liberian scene of the Keystone cops. We flew out of the doors and there stood our wives.

"They went that-away." Both of them jabbed their fingers toward the left.

Jerry and I were in hot pursuit. The three men had fled into the underground garage. We didn't have any guns with us, but I had a finger I could use. I held my hand out in the shape of a gun. They were scared to death of that finger. We searched in the dim lighting of the garage. The trio was hiding behind a pillar. I could see parts of two men. I knew the third one was there, too.

"You'd better come out." I raised my voice louder. "If you don't come out, I'm coming in to shoot all three of you." I hollered some more.

They edged out from behind the pillar.

Jerry and I quizzed them. They were from Palestine.

"Who was talking with you?" I said.

"Mr. Arafat." One of the Palestinians spoke decent English.

"Yasser Arafat?" I said.

"Yes. Great man." The same guy answered again.

"Listen." I pointed a finger at them. "If you move, I have a gun stuck in the back of my belt and I'll shoot the three of you in front of God and everybody." I didn't have a gun anywhere, but they believed me. I escorted them, along with Jerry, from there to the Ministry of Defense Building, which wasn't far away. I told Jean and Virginia to meet us there.

We arrived at the Chief of Staff's office and I prodded the three guys inside.

"Do you know these guys?" I hooked my thumb at them.

"No, I don't know them." The Chief shook his head.

"I do," I said. "They are from Palestine—Yasser Arafat's henchmen. This trio was talking on the radio, their radio, using my repeater antenna. I want them kicked out of this country right now."

"How am I going to do that?" He waved his hand at them.

"Chief, call a couple of your soldiers in here." Why was he saying, 'How are we going to do that?' I couldn't believe the Chief of Staff of Liberia didn't know how to eject people from his country.

"Chief," I said with as much patience as I could muster, "have your soldiers come in here with a truck, load the Arabs in the truck, and take them out to Sprigg's Airfield." Sprigg was the name we used for the James Sprigg Payne Airfield in Monrovia. "Book them for a trip to Spain or Europe and have them catch the first airplane out of here."

He did. The soldiers arrived and led the Arabs out of his office. One escort was a good soldier, Thomas Quiwonkpa. When you say his name, it sounds like 'Qwampa.'

"Quiwonkpa." I stepped outside of the Chief's office with the escort detail. "Keep your eye on them and don't let them get away from you. Take your gun with you."

"I got it." He patted his rifle.

"Take them under arms and put their butts on any plane leaving for Europe."

He made it happen.

I went back to the Chief of Staff.

"Let me tell you something." I stood facing him, hands on his desk. "If you want us to leave this country, you keep letting these terrorists into your country. We don't have to stay. If you don't need our American dollars, I'll make daggone sure we close this outfit down and we'll go home."

The threat hit him hard, because we provided all their military equipment and a whole lot more support.

"If my repeater station goes off air one more time," I said from the door as I was leaving, "you're going up there and get them."

The Palestinians never came back, but they left their radio. Jerry sent the radio to the CIA headquarters. They harvested all types of intelligence off the internal hard drive and took it apart. As a matter of fact, one of Jerry's agents in Liberia hand-carried the radio as a courier back to CIA headquarters. The Agency unloaded a vast amount of intelligence off that piece of equipment.

I never heard about the Arafat incident again, but Jerry said Arafat had put a hit out on both of us. I was in Liberia until the fall of 1980 and never had a problem with any threats. But when the chase and capture was all over, Jean and Virginia were laughing hysterically because it was like the Keystone cops—"They went that-away."

EUCOM VISIT TO LIBERIA AND BRIEF TO EUCOM—FALL OF 1979

Everything in Liberia was a new experience for me. Every day something was different. You never knew what was going to happen next.

I received a call from an Air Force Colonel in European Command (EUCOM), located in Stuttgart, Germany. He wanted to come to Liberia for an official visit since we fell within their military chain of command.

"You mean you want to come in an official capacity?" I said.

"Yes," he said, "that's what I want."

"Who is sending you here?"

"The Deputy EUCOM Commander," he said, "which is a Major General."

"Wait a minute." I sat straighter. "The EUCOM commander is a four-star General. Certainly his Deputy is more than a two-star."

"Well, he's an assistant to the Deputy." He backed down a bit. "A two-star Army General."

"When are you going to arrive?"

He told me his arrival date.

I explained I would have someone drive him from the airport. When he landed at the airport, I had a soldier waiting for him in an Army van. I had him brought to my house instead of sending him to a hotel. If he stayed in a hotel, he would not be happy with the experience. I wanted to give him a good reception.

While he was at my house, he didn't go around much but drank a lot of my booze.

"How many dependent children are there?" he asked.

"There are eight," I said. "They go to the American Cooperative School, where my daughter goes."

"How do they get to school?" he said.

"They ride in an Army van."

"You're using a military van to take kids to school?"

"Yes, I am." I didn't like his tone or line of questioning.

He changed the subject, asking more questions. Then he requested to have someone take him around the area to observe our operations, working locations, and Camp Schieffelin.

I had an officer escort him around Monrovia. His visit and stay in my home totaled four days. He ate most of his meals at our place.

He caught a flight back to Stuttgart. Within a week I received a call from him telling me the EUCOM Commander, the four-star General, wanted to see me ASAP (As Soon As Possible).

"What does ASAP mean?" I said. "Remember, in this part of Africa, ASAP might mean a week from now. Do you have a flight scheduled for me?"

"Oh, no." he said. "No, we don't know how to schedule foreign flights."

"In that case, I'll be there when I get there."

"You had better hurry." His tone changed to a harder sound.

"The four-star General wants you now."

The EUCOM commander was an Air Force General, which meant he might understand the difficulty of what he was asking. Or he might not.

I took a Swiss Air flight five days later. To get out of country, you have to wait until an airline flies into the airport. Not a lot of people fly into Liberia. The Swiss Air flight took me straight to Stuttgart. I arrived on Sunday and the Air Force Colonel met me at the airport.

I had packed light because I had to make an immediate flight back to Liberia. The time of this visit was when the country's internal turmoil was growing. Events in the country showed greater deterioration daily.

The Colonel drove to the two-star General, the Assistant to the Deputy EUCOM Commander, for a courtesy visit to his office. The General thought he could buffalo me.

"What took you so long to get here?" he said after I sat.

"In Africa," I said, "arranging anything takes a long time."

"What do you mean?" He drilled me with his eyes, challenging me.

"You have to wait for the airplane to get there," I said, explaining with patience. "Not many fly in and out, particularly out this way. The delay is because I had to wait for on outbound aircraft to Stuttgart."

He didn't like the answer and decided to be kind-of gruff with me.

I didn't get riled, but just gave him direct eye contact. I didn't speak anymore to him except to say yes, okay, or to answer with a shake of my head.

"We'll send you over to General Allen," he said. He was the four-star Air Force General commanding EUCOM.

"Alright," I said.

For civilians, all of this happening on a Sunday might be unusual, but senior officers in the military often had to sacrifice part of their weekends to duty requirements.

A military driver brought a car. The Air Force Colonel and I climbed in and rode to General James R. Allen's office. The General's aide was in the office's anteroom.

"COL Gosney?" He rose and shook my hand. "The General is looking forward to meeting you."

"Okay," I said.

"Come on." The aide motioned me forward. "I'll take you in."

My Air Force Colonel escort stood to walk in with me.

"No, Colonel." The aide held out a hand to stop him. "You stay out here. He doesn't want to see you."

I felt like clapping my hands.

Inside the office proper, I found General Allen to be the nicest man.

"General Allen," I said, "I got to Stuttgart as fast as I could. It took me five days, due to the availability of an airline headed this way."

"I understand." He sipped his coffee. "No big deal. Don't

worry about it."

What happened next caught me a little off guard. Events don't always unfold the way you might expect.

Before I had made the trip to Stuttgart, our Ambassador had to leave country for a medical exam. Julius Walker, who took over the U.S. Embassy as the Chargé d'affaires, wrote a letter without my knowledge to General Allen explaining my role in Liberia.

"Bob," Julius had said before I left, "are you concerned about going to EUCOM?"

"Are you kidding?" I said. "What are they going to do to me? Condemn me to serving a tour in Liberia?" We both laughed. "No, I'm not concerned."

The General had received Julius' letter by diplomatic pouch the day before I arrived at EUCOM for the scheduled visit. When the General seated me in his office, he took a chair close to me, rather than sitting behind his desk, and brought his coffee. Before I briefed him about Liberia, he reached behind himself and pulled the letter off his desk, handing it to me.

"Have you seen this?" he said.

I turned it over, inspecting it and realized Julius Walker had sent it.

"No, I haven't," I said. I read it quick and felt great. "It's the first time I've seen the letter."

"The Ambassador sure thinks highly of you."

"Yes, sir, he does." I handed the letter back. "He's been most supportive."

"Now, tell me about Liberia." He paused, waiting for me to get comfortable.

"Well, General, where do you want me to start?" I grinned.

"From the beginning."

"This might take a while."

"I've got plenty of time." He raised his coffee cup and took another sip.

I started with the 14 April 1979 uprising, the rice revolt, the underlying tension in the country, and then explained the current situation.

"Tell you what." He set the coffee cup on a coaster with a slight click. "Do you have time to have lunch with us?"

"Yes, sir." I nodded. "I do."

"What time is your flight leaving?"

"Six o'clock tonight."

He thought we had plenty of time. He called his staff. They were all flag officer rank, Generals for the Army, Air Force, and Marines, and Navy Admirals. He called a Navy Admiral, a Marine Brigadier General (BG), and a few others, five total. One was Brigadier General Blackmon who was in charge of transportation.

"The visitor we sent to Liberia, an Air Force Colonel, reported you had used military vehicles to take your dependents to school," General Allen said.

"Yes, sir," I said. "Correct. One of my officers drives the van. Having a U.S. driver was one way I could ensure our dependent kids had proper protection. When school is out, we bring them back to their homes in a military van."

"That's not any different than what we do right here," he said. "We have a military bus transport our kids to school and pick them up to take them home."

"Yes, sir." I smiled. "The similarity is striking."

The Air Force Colonel had made a big deal of our use of military vans in Liberia to transport our kids, but General Allen understood why we did it.

When we arrived at the lunch and settled in, General Allen opened the meeting by turning to me. The quickness of the flag officers' response was amazing. They arrived in no time. This was Sunday and I expected most of them to be with their families.

"I want Colonel Gosney, the Chief of the Military Mission to Liberia, to provide you an executive laydown about the information he briefed me on."

"General," I said, "where do you want me to begin?"

"Like you briefed me." He gave me a hand motion. "From the start."

"Oh, my," I said.

I finished without any questions from his staff.

"Do you see any issues," GEN Allen glanced at BG Blackmon, "with a military van taking children to school in Liberia, picking them up after school, and driving them back to their homes?"

"I see nothing wrong with that." Blackmon shook his head. "We transport our kids, as you know General, with school buses to take them to school and bring them home the same way."

After these comments, we discussed several other issues. Lunch went on forever, it seemed, but we finished lunch and General Allen got through talking with all his staff.

"Any further questions of Colonel Gosney?" His eyes traveled around the group.

"No," they said. "He covered everything."

"We needed to know more about Liberia's current situation and recent history," one officer said.

"Okay, thank you for coming." General Allen stood, which dismissed his staff.

We rode back to his office.

"Do me a favor." Allen raised a finger toward me.

"Yes, sir, whatever it is."

"Don't send a message to me," he said. "But every month or two, send me a letter by diplomatic pouch to let me know what's going on down there."

"I'll do it, sir."

"Don't send a message because I'd be about the tenth person to read it." His lips curved into a grin.

"I'll send you a letter by diplomatic pouch."

When I returned, I did what we agreed on. I'd get a nice letter back from him every now and then, which said, "I sure appreciate it. You're doing good work there."

I wanted to write him about the daggone two-star Army General who tried to put me on the defensive, but I didn't do it.

After we finished and I was leaving General Allen's office, he told me to use his driver and sedan for the rest of my visit. I didn't have much time left and I couldn't miss my flight back to Liberia. Each time I flew to Stuttgart, which wasn't often, the ladies provided me a list of things to buy at the Post Exchange (PX). I had an hour and a half before I had to catch my return flight.

"I'll take you to the airport." The Air Force Colonel stood when I came out.

"I don't need you," I dismissed him with a hand wave. "The General's driver is taking me to the airport. But let me tell you something. Don't ever set foot in Liberia again as long as I'm there."

That ticked him off. He did come to Liberia again. But I wasn't taking any guff from him and I had to get in the last punch before I left to go back.

"Take me to the PX," I said to the General's driver. "I have a long list of items these ladies want. Their husbands work for me."

I bought perfume and worked my way around the store. I flew back to Monrovia by Lufthansa. They charge the heck out of you for overweight luggage. Though I had packed light, I was overweight with all the ladies' special items. I paid an extra fifty bucks for that little suitcase.

Jean met me at the airport. Driving home she asked me about my trip.

"How did it go?" she said.

"Great."

"Julius is dying to know," she said.

"I'll call Julius when I get home." I relaxed back into the car seat.

I called Julius.

"How did the trip go?" he said.

"Excellent," I said. "The General was most appreciative that I came. I briefed his staff on Liberia's current situation and it was a pleasant visit."

CHAPTER 9
Liberia, Africa – The Coup

Liberia's history as an unstable nation, evidenced by the Rice Riots in 1979, its massive internal corruption, and the discontent of the poor created a powder keg explosion in 1980. The True Whig Party, the party of the Americo Liberians, had ruled Liberia for over one hundred years. Acting more like a dictatorial upper class, the Americo leaders had pushed their supremacy too hard. The people demanded a political landscape with more than one party, one including the Tribals. Intellectual and upper class Tribals in Monrovia did not like the elitism of the Americo population. When the enlisted coup occurred, those leaders rallied around the NCOs who took swift and dangerous action to turn the tables.

The Coup by Master Sergeant Doe—April 12, 1980

Jean, Amy, and I were in Liberia when the April 12, 1980 coup occurred. No real plan existed to pull it off, only disgruntled soldiers waiting for the best opportunity.

Seven soldiers, including MSG Kanyon Samuel Doe, lay on the beach the evening of 11 April. They had been drinking palm wine and were drunk.

"You know," one of the soldiers said, "the President is staying in the mansion tonight."

President Tolbert had a mansion in Monrovia he seldom ever stayed in. His normal residence was in the mansion up-country.

"He is?" another soldier said.

"Yeah." MSG Doe said. "My brother told me he's staying on the fourth floor."

"What are we going to do?" a third soldier asked.

"My brother is the armorer for the palace guard," Doe said, "and if we get some guns..." He motioned toward the mansion.

By this time, it was three or four o'clock in the morning on 12 April. The group weaved their way toward the compound surrounding the mansion. They climbed over a tall wall and snuck over to the Armory, where Doe's brother worked. One knocked on the Armory door. The brother slept inside as part of his duties. He didn't stir. No answer.

They banged the door.

“We need ammunition and guns,” one soldier said. “We’re going to kill the President.”

“Come in.” Doe’s brother opened the door. He issued several kinds of guns and plenty of ammunition. Their muddled minds still didn’t have a plan or clue as to how to do this right. The brother gave them more information. As the armorer he knew the President’s exact location, who was on guard, and what kind of ammo they had.

They figured out an attack. Since Doe was the senior man, he and another soldier, Wey Syen, would stay outside. When the soldiers killed the President, they were to shoot out his window as the signal to Doe and Syen that they had succeeded.

The soldiers walked right in the front door. Quiwonkpa led the attack up the stairs. The elevator to the fourth floor suite was in the lobby, but the soldiers didn’t know how to work it, which meant the safe route was via the stairs.

Brigadier General (BG) Raylee was the commander of the Palace Guards. He was a Tribal officer, a native Liberian, not an Americo Liberian. Quiwonkpa met BG Raylee on the stairs, weapon in hand.

“We don’t want to kill you because you’re a Tribal.” Quiwonkpa motioned for the General to leave. However, Raylee went for his pistol and the group killed him.

With the Palace Guards Commander shot, they hustled to the top and arrived at the President’s Residence Suite. The President’s quarters were the entire 3rd floor.

A Liberian sergeant guarded the entrance to Tolbert’s room.

When he heard the shot killing BG Raylee, he checked out the action. Because the group outnumbered him and had killed his commander, he dropped his weapon. Quiwonkpa ordered him to grab his gun, run down the stairs, and leave. The kid did as he was told. He scooped his gun off the floor and ran away.

One soldier checked the door, working the handle with quiet movements. It wasn't locked. The attack force crashed into the suite.

By the soldier's own admission, President Tolbert's state of dress made a difference. If he wore his white safari suit and had his swagger stick, which was solid gold with a lion's head on the end, Quiwonkpa and his attack team couldn't kill him because the bullets would pass through him. If he was in the bathtub they couldn't kill him because he would swirl down the drain and get away. They believed in his juju.

President Tolbert was in his pajamas and didn't have his swagger stick. They fired at him, but didn't aim well. Tolbert offered them a million dollars for his life, but the soldiers didn't know what a million dollars meant to them, so they killed him.

Quiwonkpa and the soldiers ran toward the windows, shooting them to signal Doe and Syen. The windows were bulletproof glass. Bullets ricocheted all around the room. The group shot about thirty times and the windows remained unbroken. That scared them to death. They thought the President wasn't dead yet, but using his juju—his magic powers. To ensure he was dead, they decided he had to be cut apart. They dissected him and cut his heart out, then shot the windows again. The bullets ricocheted once more. Their fear rose to a feverish pitch. Tolbert had to be in power, even though his heart was out. Frantic, they burst out from the room. At that point, Tolbert's wife made an appearance.

She had been in the other bedroom, heard the commotion, and crept into the hall.

"Go back in, Mama." Quiwonkpa said. Liberians called the women Mama. "Mama, you go back," another soldier said. "We're not going to bother you."

They raced down those stairs as fast as their legs could move.

Right behind the Executive mansion, out the back door, was a big, beautiful Chinese pagoda. The entire group of seven soldiers cowered in the shrine. Scared. They sent three soldiers to the Embassy to find me.

As this commotion was going on, an aide to the Ambassador called me.

"What's going on?" the aide said.

"Heck, I don't know." I propped myself up in bed. "I'm at my house. What do you hear?"

"A lot of gunfire." His voice sounded like he was in a small room.

"Where from?" I said.

"The palace." He stopped. "I mean the Executive Mansion."

"Where are you?"

"We're under the bed."

"Okay." I swung my feet out of bed. "I'll be there in about twenty minutes." I whipped on my uniform, got in the car, heading toward the Executive Mansion. I didn't take my driver because I left the house at about 3:00 o'clock in the morning. Driving toward the Mansion, I kept seeing soldiers running away in all directions like disturbed barracuda.

Soldiers who spotted my car coming ran to let me know soldiers had killed the President. But they don't say killed.

"Kee-oh, kee-oh." One soldier loped up to my window.

"Tolbert?" I stopped the car.

"Yeah." The soldier's head bobbed. "Tolbert. Kee-oh."

"Okay." I motioned to the rear. "Open the trunk and put all your guns inside."

"Yes, sir." The soldiers around my car followed orders.

"Now go to Barclay Training Center," I said, "and I'll meet you there." Barclay was close by, near the ocean, and housed the Executive Mansion Guard and the 3rd Infantry Battalion.

They lit out for the Barclay Training Center. I pulled away. I had so many guns in the trunk of my Ford Taurus that the back end sagged low to the ground and the front wheels skimmed the road. I felt like I was in a "low-rider" car from the seventies.

I went to the Embassy, not the Mansion, because Julius Walker, the Deputy Chief of Mission (DCOM) had radioed me. The drive didn't take long.

"What's going on?" Julius said as I walked in.

"I think they killed Tolbert." I said. "The soldiers are in charge."

About that time, four or five soldiers showed up at the Embassy gate. The Marine guards weren't letting them in.

"We wanna see Chiefo," they said. "We wanna see Chiefo."

A Marine guard contacted us and said, "They want to see the Chiefo."

Chiefo was the name the soldiers used for me. I'm not sure who started it, but it may have been because I was made an honorary tribal chief.

"Don't go, Bob," Julius said. "Don't go."

Julius Walker, the second-in-command at the U.S. Embassy, had assumed duties as the Chargé d'affaires, which meant he was acting as the Ambassador, who was home in the United States on sick leave. Julius was doing more than giving me a friend's advice. He was trying to keep me safe.

"Heck, they won't bother me." I said. "I'm going to see what's going on."

I walked out of the Embassy to the gate. The Marines were standing tall.

"Who wants to see me?" I asked the Liberian soldiers.

"Sammy," the leader of the group said.

"Sammy Doe?" I said.

"Yeah, yeah." The leader gave a vigorous nod.

"Where is he?"

"In the Chinese Pagoda." He didn't say it with those words. He told me Sammy was behind the Mansion.

"Alright, I'll go over there."

I jumped into my sedan after telling Julius my intentions.

"Be careful, Bob," Julius said. "We don't know what they'll do."

"I'm not afraid of them shooting me." I smiled. "I'm afraid of a ricochet."

I drove out of the Embassy and didn't pick up the soldiers. The Marines opened the gate and those little soldiers ran behind me 'till I reached the Mansion. The distance was maybe three quarters of a mile. I circled behind the Mansion to the Pagoda, climbed out, and strode inside.

Seven terrified young men huddled together, eyes darting around, weapons close.

They were sure there'd be a counter-coup and waited for the attack any minute. The group knew the Army's size and thought someone in the Army would kill them.

"No." I said. "There won't be a counter-coup."

"How do you know, Chiefo?" MSG Doe said.

"Because you killed the President. Who's gonna lead the counter-coup? I'll bet the rest of the leaders are hiding in the bush."

Formation of the New Liberian Government—April 12, 1980

"Julius." I spoke into the radio. "The leadership of the coup is here in this pagoda. They've got all of the government officials already under arrest and slammed them in that terrible prison here in Monrovia. The rest of the leaders and important people are in hiding. I don't think there will be any counter-coup."

"Understand," Julius said. "We need to think through the next step."

"This is going to be absolute chaos." I put on my thinking

cap. “We’ve got to form a government here. Otherwise, the coup will turn into one giant bloodbath.”

“What do you suggest? What can you do with that bunch of guys?”

“I suggest we form a government.” I made an internal decision. “I’ll appoint a government from the ones here. It’s better than nothing.”

“Do what you think is best.” Julius had me take charge.

Once the shooting started, most of the young enlisted soldiers had taken charge, arresting senior officials without any real direction from the seven soldiers who had killed President Tolbert. The enlisted coup continued of its own accord.

The seven soldiers in the Pagoda paced, acting nervous and scared.

“Sit down,” I said. “I’m going to form a government. You’re going to have a great responsibility to keep this country out of a civil war and get it organized.”

I checked them out. Sammy Doe was the senior man.

“Kanyon Doe.” I pointed. “As a Master Sergeant, you’re the senior man here. You’re the President of Liberia as of now.”

The one mistake I made was the next senior guy, Wey Syen, who turned out to be a bad guy—a monster.

“Wey Syen, you are the Vice President.”

I went by rank.

“Quiwonkpa, you are the Commanding General of the Army.”

Quiwonkpa was a sharp guy. I continued right down the line.

Then I appointed one of the soldiers the Secretary of Defense. Next on the list of officials was the finance minister position. There were three guys left. I wasn't impressed with my choices. But I asked the right question.

"Tell me," I said "what is ten times ten?"

Three heads came up and each looked at me. I inspected their expressions for comprehension.

"Do any of you guys know what ten times ten is?"

Harrison Pennue, a really short guy, threw a hand in the air.

"How much?" I said.

"One hundred." He said it with great assurance.

"You're now the Secretary of the Treasury." I nodded.

I finished making the appointments and told them to take their posts.

"Do not, I repeat, do not kill anybody yet." I tightened my jaw and gazed into each man's eyes. "And I said, 'Yet.'"

I didn't want them stormiing offices, killing Secretaries, officials, and others in this coup.

"Don't kill anybody until I tell you." I said.

I had formed the government.

Stealing from the Bank of Liberia

The Liberian Army was running amuck.

While I was finalizing the government in the Pagoda, two armed soldiers entered shoving the Bank of Liberia President before them. They also lugged a big footlocker full of American $100 bills, wrapped in bundles—stuffed to the brim. The soldiers had made the bank President load the locker full of American money. I don't know how many hundreds of thousands of dollars it was. They showed us and locked it tight.

"What the heck is this?" I said.

"I have to pay the soldiers," MSG Doe said.

"Not by robbing the Bank of Liberia, you're not," I said. "You're the President of this country. The bank is the only way you get money. You tell those soldiers to take the money back to the Bank of Liberia and don't bother the President."

"Okay, Chiefo."

"Return the money," Doe told the soldiers. "And don't hurt the President."

Next I spoke to the President. He was scared to death because the soldiers could kill him in a heartbeat.

"When you get to the bank, lock the footlocker of money in your safe."

"How will I pay them?" Doe said. "The soldiers."

"Get organized first and you'll figure out how to pay them, but you're not going to steal it out of the Bank of Liberia."

With that problem out of the way, I planned to stay with Doe for a while. He was a young man, a mere twenty-eight years old, with little background to be a President.

News of the Coup Reaches Our Family in the United States

When the coup occurred, before the Liberian soldiers took over the radio and the telephone stations, my home phone worked for a little while.

My brother, Gary, was the first to reach us on the phone.

Pam, my oldest daughter, had gotten out of bed early to prepare to do something. She turned the radio on and heard, "There's been a military coup in Liberia. All lines are down. There is no contact with anybody."

Pam panicked and called Gary, who lived nearby in Temple, Texas.

"I'll find out what's happening," Gary said.

He called Fort Hood and talked to the Chief of Staff.

"Can you tell me about the military coup in Liberia?" Gary asked.

"What?" the guy said. "Never heard of it."

"There's been one," Gary said.

"Really? Where'd you hear that?"

"My niece heard it on the radio and called me." Gary explained the details. "I have a brother, a sister-in-law, and a niece over there."

"I don't know, but I'll find out." The Chief of Staff hung up to work the issue.

Gary called the Defense Department. They didn't know what

was going on in Liberia, either. In the end, he called President Jimmy Carter's office.

"I want to talk to somebody who can tell me about a coup over in Liberia," Gary said to the operator.

"We haven't heard anything about a coup," the operator said.

"Let me talk to somebody who would know," he said.

"Let me see if I can get the Chief of Staff."

The Chief of Staff, Hamilton Jordan, came on the line.

"Did you want to know about a coup in Liberia?"

"Yeah," Gary said.

"We haven't heard anything," Jordan said.

"I can guarantee you there's a coup in Liberia."

The Chief of Staff found out there was a coup and let Gary know he was right.

"Then I'll dial my brother's phone number in Liberia," he told Jordan.

Gary made the call and Jean answered the phone. That was the first time the phone had rung in several days, but it worked well.

"Amy, put down the shotgun." That was the first thing Jean said as she was lifted the receiver. Our youngest daughter, Amy, followed the directions.

"What's going on," Gary said.

"Hours ago," Jean said, "we had a coup here and Bob's with the troops trying to get it under control."

"Are you all alright?"

"Yes," she said. "We're fine. They won't hurt us. They're after the other people. We're not concerned. Bob told me, 'If those Liberian solders come to our gate, you take the shotgun and shoot it one time over their heads and tell them, "Next time I'm shooting you."'"

We had a big wall topped with barbed wire extending around our compound, like in Mexico. The entryway was a huge iron gate in front.

Jean did what I said. The soldiers ran and drove around our home. To drive to the house, you took a dirt road a quarter of a mile from the one paved road in Liberia.

I was heading home and Jean radioed. She told me people were outside the compound hollering and shooting in the air.

"Do what I told you." I said. "Get the shotgun and go out there. Open the door and tell them, 'You step through the gate and I'll shoot.'"

And BAM—she shot one over their heads. They sprinted away, hollering. About halfway down the dirt road, they stopped me.

"Don't go home, Chiefo," a Sergeant said. "Don't go home. Missy vexed." I laughed. He meant she was ticked off. Those events were a comedy of sorts.

Pam called after Gary got through. She described the news broadcast about the coup in Liberia.

"Don't worry," I said to her, "we're alright. We're not concerned at all."

The phone went dead.

That's because the coup boys whose job was to shut down the phone and radio systems finally made it happen. I didn't have any communications back to the States.

But in one of the quarters on post, I learned a CIA official had a shortwave radio with a big antenna. He talked to contacts in the United States and also let the Embassy people talk with their families.

I used the radio to tell Gary our status.

"Everything's okay." I said. "Don't worry about us." I could hear the message passed to him. The radio operator did a relay in the States, who called on a phone. That's how I kept my family notified we weren't in any danger.

Executions After The Coup —Tuesday, April 22, 1980

On Monday, April 21, 1980, nine days after the coup, I went to the Liberians' main Army camp, Camp Barclay. The beach is right next to it. I saw a telephone truck drive onto the beach. The men drilled holes in the sand and erected four telephone poles. I thought, "They're going to have a big execution here. Soldiers will tie the accused to poles and shoot them."

I reversed course and visited the Embassy to talk with Bob Smith, the Ambassador, who had returned from sick leave.

"If they execute anyone," I said, "I bet it'll be no later than tomorrow."

"You think so?" he said.

"Ambassador, I know so."

The next morning I drove to the Camp. The place had filled with a sea of people like barracuda jammed together. Ten thousand Tribals with machetes, spears, and costumes swarmed the area like barracuda waiting for a kill.

I slowed, maneuvered past the gate and through the crowd. They parted. I moved to the beach and found Quiwonkpa, the new head of the Liberian Army.

Quiwonkpa had a firing squad. Two mini-buses muscled their way to a rise by the beach. Inside the vehicles were thirteen governmental officials they had tried: the Secretary of Defense, the Attorney General, and several other agency heads from Tolbert's staff. In the vans they could see the poles on the beach. I knew nothing could be done because the barracuda demanded the new government kill these people.

"If I don't execute them," Quiwonkpa said, "they'll kill me."

Soldiers pulled four condemned officials from the vans and forced them to march to the poles. Quiwonkpa prepared the firing squad.

They aimed. The crack of rifles jarred the crowd to momentary silence.

I couldn't believe it. The squad didn't hit a vital part in any of the officials. The shots hit their legs, shoulders, or another body part. I made Quiwonkpa go down and do the coup-de-grâce. He shot them in the head.

"Bring your firing squad over here," I said.

We walked out to a fifty-five-gallon barrel sitting a little bit away from the telephone poles.

"Alright," I said, "have them shoot that barrel."

They missed the daggone barrel. Forty feet from the target and they missed it.

"Move closer." I waved them forward.

They fired and missed again. I had them get about ten feet from it before they fired and hit the barrel, covering the entire target.

"You get ten feet from those officials with the firing squad," I said, "and shoot them one at a time. Don't try to shoot four at once, because you won't kill them. You'll make them suffer."

They led four more to the beach. Can you imagine sitting on the bus and seeing this? The soldiers had the condemned men's hands tied behind them, got them to the beach, and tied them to the poles with ropes under their arms and around their waists. The firing squad got ten feet out and shot them, first one, then the next one, until they finished. The soldiers cut them loose and brought the next group.

An old official, the convicted head of the Supreme Court, had a heart attack. He died before the soldiers got him to a pole. The squad went back and got another one. After they had executed them all, the dump truck backed into the area. Soldiers threw the bodies into truck's bed and hauled them off.

Quiwonkpa told me they dug a hole with the electric company's backhoe, dumped the bodies inside, and covered it.

I left for the Embassy and reported the events to the Ambassador. He wasn't shocked; he was disappointed.

"I wish you had seen it," I said. "I have never experienced anything like that. At least ten thousand Tribals gathered there like a school of killer barracuda."

THE DEMARCHE

Ambassador Smith called me to his office after the coup and his return.

"Bob," he said, "come to the Embassy and go with me to the mansion. I've got to give a demarche to Sergeant Doe."

A demarche is a formal appeal or protest written in diplomatic language from one country to another, from our head of state to another, or vice versa. When giving a demarche, you are obligated to notify the State Department. I joined my boss and we went to the Mansion to talk to Doe. The Ambassador delivered the demarche, which in this case was from our President to the President of Liberia.

Afterwards, we drove back to the Embassy.

"Bob, how do you think I did?" he said. "Do you think he'll follow those instructions?"

"He didn't understand a dang word you were talking about," I said.

"Are you kidding?" he said.

"Did you see him? You talked about thirty feet over his head. You should have said in plain English, '...don't kill anybody or we're going to withdraw all financial aid to your country.'"

"Can you go back, talk to him, and let him know what I said? What I meant?"

"I sure can," I said.

"Drop me off," he told his driver, "and take the Colonel back over there."

The driver took me to Doe's big Presidential office in the Mansion.

"Sammy," I said, "let me sit here and tell you something. I know you didn't understand what the Ambassador said."

"No," he said. "No, I didn't."

"The Ambassador said, 'If you start killing these ex-governmental heads, they're going to withdraw all the United States aid you get, which is about a hundred million dollars a year.'" I shrugged. "You won't get that."

"Okay." But he said, "I don't know if I can manage without it."

"Do the best you can because you're going to lose a hundred million dollars."

"Can I kill a few of them?" he said.

"It's up to you." I shook my head. "But I can't guarantee you'll get any money. The Ambassador said if you kill any more of these people, you'll lose all support from the United States which you have had for 140 years. That's what he wanted to tell you."

"What do you think, Chiefo?" Doe said.

"Losing that support is bad." I shifted in my seat. "You don't need to kill any more of these people."

"But there's two of them." He named them off.

"Two more?" I thought for a second. "Okay, but no more."

One of them was Tolbert's son, who was hiding in the French Embassy and the other one was Tolbert's other son. They were both big crooks.

President Doe stopped killing any more people after he got those two. But he hadn't understood a single word of what Bob Smith had said.

The Mercedes Debacle

The leaders of the coup saw forty-eight brand new Mercedes automobiles in the parking lot. They wanted to know where the keys were to the cars.

"I don't know," I said. "Does anyone know how to hotwire a Mercedes?"

"No," they said.

A military officer whose line of work was in the Ordinance Corps worked for me. I called him.

"Come to the Mansion," I said. "Do you know how to hotwire a car?"

"Yes," he said. "I do."

"Show these guys how to get these cars running."

He helped them hotwire all forty-eight Mercedes. Within three months the drivers had crashed every last one of them.

Jean and I were coming home from the American Embassy and my driver was driving. In a flash, a Mercedes pulled beside me with one of the soldiers at the wheel.

"Chiefo, Chiefo," he shouted, pointing at the car. He had to let me know he was driving a Mercedes. Speeding ahead, he had a head-on collision about half-of-a-block in front of me. He smashed into a taxi.

"Don't stop," I said to the driver. "Keep going."

Two days later the soldier limped into my office. His arm was broken, he had a cast on his arm, and he had cuts all over him.

"You see me, Chiefo?" he said, chest puffed up with pride.

"Yeah. I saw you." I nodded. "What did you run into?"

"A taxi cab," he said. His voice lowered. "The car is spoiled."

"Alright."

He left moments later, another walking example of the lifestyle in Liberia.

President Tolbert's Son

One Liberian, Tolbert's son, was a worthless criminal. He had broken into the French Ambassador's house and demanded asylum. He didn't knock on the door; he forced his way inside. He scared the wits out of Ambassador de Low and Madam de Low. The Ambassador came over to the American Embassy. I was upstairs talking with Bob Smith, the American Ambassador, and de Low said he was frightened to death.

"I've got Junior," he said, "Tolbert, Jr. living in my residence." The Ambassador and his wife lived in the French Embassy in an apartment upstairs from the offices.

"What do I do?" he said.

"If you can keep his mouth shut," I said, "we might get him out of the country."

But Tolbert, Jr. was a loudmouth. He was an obnoxious scoundrel, close to thirty-five years old.

"Tell him to lay low and don't made a scene," I said.

The Ambassador came over the next day.

"He's on the phone calling people in Liberia." He was distraught.

"I'll tell you what's going to happen," I said. "You two get upstairs and lock yourselves in the apartment because Doe's soldiers are going to come get him."

The soldiers broke into the Embassy. Now, a soldier breaking into an Embassy is an act of war. But Ambassador de Low didn't care at that point.

They dragged Tolbert out of the French Embassy, drove him out into the countryside and killed him. The troops dumped him on the side of an old country road. Tolbert's son was the last killing Doe's forces did.

SECURITY OFFICER

The American Embassy's Assistant to the Chief of Security kept hounding me.

"Colonel," he said, "I'm warning you. Each day you come to work, take a different route to get here."

"Impossible," I said.

"Why?"

"I have one way to get here," I said. "There are no bypass

roads. There is one road. If I'm going to come to my office, I have to drive on that one road. I can't go through the swamp. I'm not worried about it."

The Assistant had told Jerry Richardson, the head of the CIA, the same thing.

"Well?" Jerry asked me.

"For me to get to the Embassy there is one way, my dirt road, and I can't find an alternative route. Don't worry about it, we'll take care of ourselves."

President Doe's Adjustment to His New Position

Ambassador Bob Smith and I went over to see MSG Doe about three or four weeks after he became President. He sat in his comfortable office chair wearing a pair of camouflage pants and a T-shirt. His feet were on top of his desk.

When I walked in he dropped his feet, stood, and saluted me.

"Don't salute me," I said. "You're the President of Liberia. You do not salute me now."

He looked at me, puzzled, his head cocked a little. He must have been thinking, "Why not?"

I had to hold back a chuckle. He didn't get it. Though he was a MSG, he was the President of a country. Even if he had been trained by my Special Forces as a MSG and had been subordinate to me, circumstances were different now. A President doesn't salute a Colonel because he is well below the President's rank in protocol.

Soldiers Carry Guns into an American Embassy Party

Later, after Doe carried out his successful coup, while Jimmy Carter was the President, the Ambassador threw a big party at his American Embassy residence. A large amount of dignitaries and Liberian big wigs attended, those who had survived the coup and hadn't been killed. The Liberians were not high government officials, but influential people. Jean and I were there along with the Ambassador's wife and a crowd of other wives. Not many men attended.

Without warning, Wey Syen popped in with eight soldiers carrying their guns. They breezed in like they owned the place, but they hadn't been invited. An American from the State Department, Zig (Zee) Brzezinski, the U.S. President's National Security Advisor, had joined the party earlier that evening.

I wasn't paying attention to the soldiers. I was speaking with a couple Liberian men in another part of the room. Brzezinski came over to me.

"Bob, can you get those soldiers to put those guns away?" he said.

"Where are they?"

"In the room over there." He motioned toward their location.

I drifted over to the soldiers.

"Bring your men over here," I said to Wey Syen, "and follow me." I led them to a hall coat closet and opened the door. "Put your guns, knives, bayonets, and any other weapons in here because you're scaring the ladies. They aren't soldiers like you and me. They're frightened of guns. Soldiers shouldn't scare ladies. When you leave, stop by and take your gear out of the closet."

They obeyed.

One of the soldiers, Harrison Pennue, was a little guy about four feet tall. One of the waiters passed by him hefting an extra-large tray of close to fifty deviled eggs. Pennue grabbed the tray from the waiter, set it on a table, and scarfed down the eggs. The Liberian waiter looked at me and rolled his eyes. Pennue ate most of those eggs, but before he finished four or five soldiers joined him. They ate every single egg.

"Act like you've been here before." I cautioned the soldiers. "Don't do anything crazy like eating a whole tray of stuffed eggs. Be quiet and behave like gentlemen."

They settled down.

Another fellow, Richard Moose, who went by Dick, was the Assistant Secretary of State and head of the African Division.

Dick was flabbergasted the soldiers did exactly what I told them to do. He navigated through the crowd over to me.

"How did you do that?" he said.

"They respect my command authority."

"It' hard to believe they did what you told them to the last detail." He shook his head.

"They're good soldiers," I said. "They do stupid things at times like coming in here with guns, but they're hidden now."

Dick was impressed. He went outside with the U.S. dignitaries involved with Liberia to talk about their concerns and socialize.

"See the control he has over these guys?" one of the diplomats said.

"Yes." Dick nodded. "I saw it in action, but I've never seen anything like this. Colonel Gosney never raised his voice. No fuss. He said, 'Take your guns and put them in the closet.' And the solders obeyed." Then Dick told the Ambassador, "Gosney's influence over the Liberian soldiers is incredible."

"For heaven sakes," the Ambassador said, "have him stay longer because without him, there's no telling what they'll do."

In no time, Dick had come back over to me.

"I'd like to visit with you tomorrow morning at the Embassy," he said.

"Alright. What time are you going to be there?"

"Ten o'clock in the Ambassador's office."

The next day, I arrived at the Ambassador's Embassy office at ten o'clock.

"How do you feel about staying here another year?" Dick Moose asked.

"Well I don't feel too good about it, to tell you the truth." I looked him in the eye. "I've almost completed my tour and should be going home in a couple of months."

"We can make it three or four months," Moose said. "Jimmy Carter asked you to stay."

"I'll do it more for you if you ask me to stay, instead of President Carter."

"Then I'm begging you to remain here." Moose leaned forward. "You've got these people under control. Outstanding."

"Bob," said Ambassador Bob Smith, "you would do a great

service to this country if you extend a little longer."

"If the Secretary of Defense asks the Department of the Army to keep me here a few more months and they agree, I won't mind," I said. "I can't make the decision on my own. I'm due to rotate."

The Ambassador and Dick Moose called the Secretary of Defense, who called the Secretary of the Army and got his approval.

"The Secretary of the Army said, 'Tell him to stay. It's okay.'" Dick Moose smiled when he told me the news.

The end result was I remained in Liberia and became the "babysitter" of the Liberian Army for an extra five months.

That's why my tour there was longer than two years.

At the time of the party, though, I didn't know Dick Moose was an entrenched liberal, but when I taught him about Liberia's corruption he became more conservative. He visited with me a lot when he was in Liberia, wanting to know what was going on and how things worked.

I told him the truth.

"These people over here in Liberia, the peons" I said, "are good people, but the government and leaders are stealing them blind. I educated Dick about my efforts to eradicate corruption in the soldier's pay and the terrific payoff for soldiers and the country.

Vice President Thomas Syen and The Plywood Factory

My biggest mistake in naming who would run the Liberian government was appointing Thomas Wey Syen as the Vice President. He was a crazy and sadistic creep.

Greenville, Liberia had a giant plywood manufacturing plant. Most of the thousand plus workers were Liberians, but plenty were Philipinos. The plant made a major part of the country's plywood supply. A small port and airfield serviced the factory. The company owners, Americans, had their own private airplane.

One of the owners had come to Monrovia right after the coup.

"I'll meet you at the Embassy," I said to the American. He had brought a million dollars in a suitcase to pay his taxes to the government. "No one here has the authority to credit the money to your taxes."

"Really?" He raised his eyebrows.

"Don't pay it," I said. "Forget about it. The people who know about the taxes are in jail. These people will tuck the money in their pockets. Go put it back in your safe."

Three or four months later, the Ambassador got a call from one of these owners in Greenville. He called me and asked me to check it out.

"Major General Syen has commandeered an airline," the owner had said to the Ambassador, "and is trying to rob us. If we don't hand over the money, he's going to shoot us." The caller had talked as fast as he could.

Wey Syen hijacked a Liberian Airlines plane, flew to the Greenville airstrip, and walked into the plywood factory to rob the owners. He had an AK-47 and demanded the Americans open the safe and give him the contents.

The Embassy had a twin-engine King Air airplane. Soon after I had been called, I piloted the plane to Greenville airstrip by the factory. I disembarked and strode over into the owner's office area.

Wey Syen stood there, talking loud, and brandishing a gun at the two owners. The two pilots from the hijacked airline were also there. Syen's eyes opened wide with surprise when he saw me.

I didn't slow down, but snatched the AK-47 out of Syen's hand and slapped his face as hard as I could hit him. He dropped.

"Take your airplane back to Monrovia," I told the Liberian airline pilots.

"Get up." I motioned the AK-47 as I spoke to Wey Syen. "Start walking to Monrovia. Head out."

"No, Chiefo," he said, climbing to his feet. He rubbed the side of his face.

"Do you want to die or do you want to walk?" I swung the gun toward him.

"That's a long walk," he said.

I pointed my finger at his nose.

"If you come back here, I'm going to kill you." I had to threaten him to get him moving. It's all he understood. I motioned him outside. "Get your tail out of here."

He left on foot. I guess he hitched a ride because when I returned he had made it to Monrovia.

I ensured the Liberian pilots had flown away and Wey Syen didn't sneak back into the plant.

"He would have killed you for the money." I sat down, facing the owners. "What will it do to you if you lose your company?"

"It won't do anything to us," they said. They had paid the infrastructure and building bills. The owners built the aged and dilapidated structure about fifteen years ago.

"Don't spend another night here," I said. "I strongly recommend you take the money out of the safe, close this plant, and get out of this country. The day will come when I can't get here fast enough and these crazy Liberian soldiers will kill you. I can't guarantee your lives."

"We'll do that," they said. "We'll leave."

They took thirty million dollars from the safe. About a thousand employees lost their jobs. The owners left that night in their private plane.

I flew back to Monrovia and went to President Doe's office.

"If you don't get rid of Wey Syen," I said, "he's going to cause you grave problems. He'll ruin your coup. He tried to rob the plywood company."

When Wey Syen arrived back in town the next day, he tried to come in the Presidential Mansion. In spite of the fact that he was the Vice President, Doe's men took him in the jungle and one of Doe's bodyguards shot and killed him.

"Get rid of him," I had told Doe. He took it to heart. I didn't mean to kill him; however, they took my message that way.

Liberia was lots of fun.

CHAPTER 10
Liberia, Africa – Finishing Strong

Without the support of my friends in Liberia, which included bosses as well as a limited number of good Liberian diplomats, Army Officers and civilians, I would not have accomplished the great progress we made in Liberia. After the coup, the United States was still on speaking terms with Liberia and had maintained its presence as a strategic nation in Africa. We did not lose any American lives in the Rice Riots in 1979 or during the 1980 coup in the staff or military assigned to Liberia.

My wife, Jean, did a marvelous job. The Department of State awarded her the Meritorious Honor Award in August 1980 for the work she did as the consulate assistant from right after the coup until we departed. The woman who had previously occupied the position had to leave country with her husband because she was white and he was black. The Liberians jailed him during the coup and that was enough. The couple left Liberia the minute he gained freedom. Jean walked in, picked up the pieces, and did a superb job completing her mission in a manner beyond the call of duty in areas outside her assigned responsibilities. That's enough to make any husband beam with pride.

Our American Friends

A few of the American officials involved with Liberia were friends.

Julius Walker, the American Embassy's Deputy Chief of Mission for Liberia, was a classmate of mine at the National War College. Julius was from Plainview, TX.

Bob Smith, the Ambassador to Liberia, was from Belton, TX. He lived around the corner from Jean and I. He came to visit us a couple of times. Julius did, too.

I received many letters of commendation in my life, not to mention ones from Julius Walker and Bob Smith. In addition to those two letters, two others for service in Liberia stand out in my mind as well.

Zbigniew Brzezinski, the National Security Advisor to President Jimmy Carter visited us after the coup took place. He sent a Letter of Commendation through my long chain of command for the outstanding work I did as Chief of the Military Mission to Liberia. He highlighted my role during Liberia's coup stating I had "played an extraordinary and perhaps decisive role in maintaining internal stability in the new military government, thereby contributing directly to the national security interests of the United States." He also lauded my willingness to disrupt my personal and professional life by remaining for an extended tour of duty until the new regime of President Doe had settled into power.

The Honorable Richard Moose, Assistant Secretary of State for African Affairs, became a friend. He also wrote a Letter of Commendation to me.

He gave me his "personal praise and appreciation for the outstanding manner in which I performed my duties," with a special emphasis on the personal rapport I had established with the new Liberian Government officials. My rapport led to "avoiding significant damage to the United States national interests in Liberia" and my quick interventions were "very helpful in ensuring the welfare of our over 3,500 American citizens and $350 million dollars worth of American investment in Liberia."

That letter was significant to me. In the beginning of our relationship Dick had told me he was an ultra-liberal, which put us at odds, because he thought the U.S. military personnel were a bunch of ragbags. However, he was a great guy and soon changed his mind about the military due to what he saw in Liberia.

Liberian Coup Leaders Around President Doe

The Liberian personalities were varied.

Before President Tolbert took over, President Tubman was

in power over Liberia for twenty-seven years. He built a stable economy, though it was full of graft. When President Tubman died of natural causes, his wife moved to a penthouse in New York, and lives there today as far as I know. Tolbert tried to keep the country stable, but his political ambitions and greed wrecked the country's infrastructure.

I knew every one of the soldiers in power after the coup, such as Samuel Kanyon Doe, Thomas Quiwonkpa, Ben Harrison, Harrison Pennue, Jacob Harris, and others. These soldiers used what they call their Christian names. For instance, the sergeant who worked for us at our house had a Liberian name—Glahu Penau. But everyone called him Sergeant Alfred.

Thomas Wey Syen, the Vice President, was the first of the seven coup leaders killed by President Doe.

Thomas J. Quiwonkpa was the best soldier of the seven who led the coup against Tolbert. I made him Commanding General of the Army because he was a sharp guy. President Doe killed Quiwonkpa because he was a talented, smart man. Doe was afraid of him and saw him as a threat.

Harrison Pennue was a bright, wiry soldier who I made Minister of Finance because he could add ten plus ten.

His shenanigans made me laugh at times. I was in my office and heard gunfire one hot day. When I glanced out the window, I saw Pennue working with ten to fifteen Liberian soldiers in an open space. Talking loud, almost at his top volume, he tried to get them in formation. He shot his .45 caliber pistol in the air to get their attention. I got so tickled I stayed glued by the window watching the comedy unfold.

Over time, Doe got suspicious of Pennue and killed him as well.

Doe also dispatched CPT Jacob Swen, CPT Harris Johnson, and CPT William Gould. He killed anyone who looked like he might be a rival to him. Doe felt he had the right to eliminate people who might stir a revolution against him.

Rufus M. Darpoh wrote the following in his article Our Eyes Are Open, Outlook Magazine, September 1980. "Someone has written that political assassination is not murder if it is done for a genuine cause for the good of society." His article focused on the assassination of Tolbert for the good of the country, but Doe used this reasoning for knocking off anyone that might threaten his power.

Liberian Coup in the News

The coup was in the news around the world. Even the Washington Post carried a story on page A27, Liberians Bury President in Mass Graves with Aides. The article mentions me working out the government's "new security plans with Liberian Defense Minister, Major Samuel Pearson" for President Doe. Next to the article is one discussing the Liberian Envoy to the United States carrying on his business as usual throughout the coup transition. Though no Washington Post reporter lived in Liberia, a stringer, maybe the one Toney ran out of the El Meson restaurant, sold that information to the Post.

The Liberian papers, a local source for intelligence, and their reporters don't meet the same standards kept by the United States press. Articles in Liberian papers slanted or colored the news to make the new President, his staff, and the ruling party, the People's Redemption Council (PRC), look good. The PRC took over the government from the one party rule of the True Whig Party, which was based on American ancestry for membership.

A glance at any Liberian newspaper revealed immediate flaws to the eye, indicating the substantive flaws in the stories. Publishers mixed different fonts together, creating newsprint where letters appear to rise or fall half a line in several places. Editing for spelling and other errors was about 95% effective.

Coup Arrests and Burleigh Holder

The arrests of Tolbert government officials took place during the coup itself and for days afterwards. The PRC arrested, then released Franklin Smith, the Chief of Staff in Tolbert's government. To be fair to the new regime, the PRC didn't put everyone arrested through a trial. Doe's government made Smith an Ambassador to one of the African countries. That demonstrates the PRC didn't kill everyone they arrested.

Burleigh Holder, the Minister of National Security for President Tolbert, was also Tolbert's son-in-law. Burleigh was a big man and a good-natured guy. But he flew into Switzerland with trunks full of money to deposit in his account in a Swiss bank. He was one of the biggest thieves in Tolbert's administration.

The PRC sent him to a jungle prison called Bela Yala. He went there instead of being tied to a telephone pole and executed because the PRC and the soldiers in the military respected him. At Bela Yala the prison wardens didn't erect any fences to keep prisoners inside. If a prisoner escaped, they were surrounded by nothing except jungle. But Burleigh bounced back and got an eventual release.

The PRC executed the less fortunate of Tolbert's government. There are pictures of the executions in two Liberian newspapers I have saved: the Weekend News, 26 April 1980 and The Sunday

Express, 27 April 1980. I've saved the newspapers since they contain the names and pictures of those I knew who were in power one day and whom the PRC ousted the next.

I couldn't defend the executed officials or keep them from being shot because a few were pretty bad news. Testimony wouldn't have helped anyway as the tribunal was a farce. The convictions and executions were a foregone conclusion.

The single person I am aware of whom the tribunal tried and didn't kill was Burleigh Holder.

Here's a funny story about Burleigh. He drove in front of my house one day and Amy, our daughter, was outside with her two dogs. She loved those two Liberian dogs and they loved her. When Burleigh pulled in, he stepped out and those dogs barked at him, causing a huge racket.

I looked out my window and Burleigh was on the top of his car.

Liberians are scared of dogs.

"They're not going to hurt you." I said, smiling big and chuckling.

Final Advice to President Doe

Six months after the coup, I went to the Ambassador's office.

"There isn't a heck of lot more I can do here," I said. "If I leave and the country decays into chaos, I've done everything I can do to prevent that from happening."

The Liberian Army was functioning well. Their government was running real smooth.

"Don't steal money," I said in one of my last meetings with President Doe. "All the other leaders, your predecessors, got filthy rich stealing money from the people and the soldiers. Don't do it. If you do, you won't be President long because your Army will turn against you."

He didn't understand that. He acted like the rest of the leaders before him. He stole money right and left.

He dreamed of going to America. I attempted to dissuade him, but many months later, he flew to the United States.

"I want to go to New York City," he had said to me before I left Liberia. He talked about it often.

"Where do you live?" I asked him. "You live at Bosra, don't you?" That was a long distance into the Liberian up-country.

"Yes," he said.

"How long does it take you to walk from Bosra to Monrovia?" I motioned a great distance with my arm.

"A week."

"Do you know how long it would take you to walk from here to New York City?"

"No." He shook is head.

"Six months," I said. "It would take that long to walk from here to New York City."

"Far, far," he said.

The best way Liberians understood distance was by the time it took to walk to the destination. His "far, far" comment made me stifle a laugh. But he was persistent.

"I'll bring a map to you," I said with exasperation one day. "I'll post it on your wall and show you New York City's location."

Two or three days later I found a National Geographic map at the Embassy. I took it to Doe's office, hung it on his wall, and pointed as I talked him through the locations.

"That's New York City." I touched the U.S. Eastern Seaboard. "There's Africa and there's Monrovia."

I also wanted to make sure he knew a comparative distance he understood.

"Come here, Sammy." I waved him close to the map. "This dot is Monrovia. Your village is forty miles up-country from Monrovia in Bosra."

"Yes?" He looked at me for more explanation.

"We've gone over this before, but how long does it take you to walk from your village to Monrovia?"

"Oh. Five or six days by foot."

"Okay." I slid my finger along the map. "If you get to Monrovia and you want to go to New York City, you'll be walking five years to make it to New York City."

"Far, far." He shook his head.

"Yeah," I said. "A long way."

President Doe Visits President Reagan in Washington DC

After I completed the final four months of my original tour, which had been extended to a total of six months, my family and I flew home to Belton, close to Fort Hood, Texas. Jean and I decided to take a little vacation in Acapulco, Mexico.

In the midst of our relaxing vacation, the American Embassy in Liberia called Fort Hood and spoke with a Major.

"Where can we locate COL Gosney?"

"He's on leave," the Major said. "Let me call around. I know his brother lives over in Temple. I'll call him and see where the COL is located."

The Major lived in Belton. I didn't know him, but he knew about Gary and me. He reached out to my brother.

"Bob and Jean are in Acapulco," Gary said to the Major.

The next thing I know, I got a call from the American Embassy in Mexico.

"We're going to come over to Acapulco to pick you up and fly you back to Temple, Texas," the Embassy representative said. "You gather your uniforms and we'll fly you to Washington, D.C. by the day after tomorrow."

"What in the blue blazes am I going to do in Washington, D.C.?" I kept myself from raising my voice too much.

"You're going to be a babysitter for Master Sergeant Doe."

"Good grief," I said. "Is he in D.C.?"

"Yes." The Mexican Embassy rep paused. "He's on an official visit."

"Incredible." I shook my head. "I guess he made up his mind and did it."

The Embassy officer met Jean and I in Acapulco and flew us to the Temple airport. They had a driver meet us and take us to our home in Belton. I collected my uniforms, making sure I had all my dress uniforms, including my dress whites. I would have to attend a formal dinner.

In an hour I was heading to Washington, D.C. Jean didn't go because I already knew it wouldn't be appropriate for her to be there. I was thinking a hundred miles-an-hour. "What the heck is Doe doing in D.C.?"

When I arrived, the daggone State Department, which can be as stupid as a third-grader trying to be a college student, had the head director for the dinner contact me.

"You're just in time for the State dinner tonight," the director said.

"Are you going to have all of the Liberian party accompanying Doe present?" I kept my cool. "I bet he brought a whole airplane full of strap-hangers."

"Yes," he said. "As a matter of fact, he brought a number of guests. I believe it was fourteen soldiers."

"I bet he had six or seven women on there."

"Yes, he did. He brought fourteen women."

"Those women are not his wives," I said. "Make sure they don't come to the dinner."

"I was told the women who came along were flight attendants."

"They aren't flight attendants, either." I sighed. "Who is going to prepare the meal?"

"We do have a chef here." He sounded like he was getting testy on me.

"I need to talk to the head chef," I said.

I went to see the guy who was preparing the meals for the banquet.

"Listen," I said to the chef. "Tonight you'll have about fourteen men that have eaten with a spoon all their life. They eat rice with chopped chicken in it and a lot of pepper or hot, hot peppers. They won't eat anything else. The situation will be embarrassing to the President because the soldiers won't have the foggiest idea how to sit at a formal dinner with umpteen knives, forks, spoons and other special silverware. And when you bring a plate of one food and take it away, then bring a plate of another food and take it away, and follow that with the main course, they won't know what in the world to do."

"I'll remember that," he said.

"And you had better have a big spoon beside their plates," I said, "because they eat in Liberia with a big spoon."

"Oh, my goodness." He began to protest. "The table setters can't do that. President Reagan is going to be here with them."

"Oh, Lord," I said. I knew the food was going to make the event a social disaster unless I did quick, on-the-spot thinking at the dinner.

Well, earlier in the morning about ten or eleven o'clock,

Julius Walker, who was now the U.S. Ambassador to Liberia, had gone to see President Reagan.

"I need to meet this African President," Reagan said. "Can you bring him in?"

"Yes, sir, Mr. President," Julius said. "Yes, sir, I can."

Julius was escorting Doe at the time. He found Doe.

"President Reagan wants to visit with you now." He took President Doe into the Oval Office.

Doe looked around the room. He saw President Reagan sitting, waiting to speak to him.

"President Doe," Reagan said, "it's a pleasure to meet you." He stood, waiting for Doe to come over and offer to shake his hand or greet him in another fashion. But that wasn't Doe's response.

"My eyes no see the Chiefo." Doe's eyes continued to wander around the room, searching for me.

"I almost passed out," Julius jokingly told me later. "It was the funniest thing I've ever seen in my life. Doe came in to meet with President Reagan and he says, 'My eyes no see the Chiefo.'"

Ronald Reagan thought, "Good grief, I'm the Chief."

"Sir, what did you say?" Reagan waited for Doe with patience.

"My eyes no see the Chiefo." He repeated his first line.

"What's he talking about?" Ronald Reagan said to Julius.

"Well, sir," Julius had told the President, "the Chiefo is President Doe's confidant, the man he depends on. He's referring to Colonel Gosney. He's the Chiefo to President Doe."

"Where in the heck is this Colonel?" Reagan said.

"He's at a hotel right now. He arrived minutes ago." Julius moved closer to the President.

"Well, for God's sake, bring the Chiefo in."

Julius had President Doe led out of the Oval Office and talked to Reagan in private.

"Sir, remember, they're all a brick-load shy."

"I understand." President Reagan gave him a nod and dismissed Julius.

It didn't take long for me to get a phone call.

"Get over here to the White House," Julius said. "The President wants to see you."

With little time to prep, I threw on a uniform, caught a cab, and rushed over to the White House. Doe wasn't in President Reagan's office. Julius had someone escort the Liberian around the White House grounds.

I went into the President's office with Julius.

"Are you the Chiefo?" Reagan said.

"Yes, Mr. President," I said. "That's what they call me."

"Chiefo, for God's sake, take care of the Liberian President."

"Sir, I've been babysitting him for a long time."

"Get him ready for the State dinner."

"Mr. President," I said. "I need to advise you that when the Liberians sit at this State dinner, they will not comprehend what to do with the fancy silverware, napkins, and other formal table settings. The single utensil they eat with in Liberia is a big spoon."

"Oh, my." He shook his head and then faced Julius.

"If you had told me this, Ambassador Walker, I would not have had a formal State dining party."

"Don't bring any women to this place," I said to Doe after I had left the President. "It's for men alone."

I arrived at the evening State dinner to see a huge table with all the silverware I had said Liberians wouldn't use. They set it out in the typical State Department way. They hadn't listened to a word I had told them. I grabbed a menu, perused it, and called a server over.

"You stay here and watch for what the Liberians need." I moved on to check out the rest of the room.

The food had already been served on plates, which were carried to the guests. The waiters gave Presidents Reagan and Doe their plates. The protocol personnel had seated me about three seats from Doe's place of honor. Doe said something to Julius I didn't hear, but when he told me afterward, I chuckled.

"My eyes want to see the Chiefo sit here." Doe looked in my direction.

"Sir," Julius said to President Reagan, "Doe wants the Chiefo to sit next to him."

"Tell the Colonel to move forward and sit next to President Doe," Reagan said.

Protocol had to bump a couple of people down a seat to

accommodate that move. In minutes, I was sitting next to President Doe.

"Now remember," I said, "this is the President of the United States. He's the big Chief. Be sure to be nice."

At the beginning of the affair, formality required certain toasts—a toast to one thing and then another. The Liberian soldiers didn't know what to do with the toasts. Their eyes were glued on me. I'd take the glass of wine, lift it as a toast, and take a small sip. They followed my every move, took a drink of the wine, and set their glasses on the table.

The wait-staff served all the plates of food. Several waiters made a simultaneous entrance and served the meals. It was a delicious meal, made of three courses. I glanced at the Liberian soldiers Doe had in his party and they stared straight back at me. I found the biggest spoon I could find in all the silverware and I held it high. They all lifted their spoons and started to eat.

Using the big spoon went well for the Liberians because they had never eaten with anything else. The guests of Ronald Reagan couldn't believe the Liberians had eaten their food with a spoon. It didn't follow formal etiquette procedures.

However, the soldiers gagged on the food. I forget what type of meal the wait-staff brought, but it didn't have any rice or hot peppers with it. The soldiers tried to eat parts of the meat or vegetables, but it wasn't working. The scene reminded me of a disaster. Julius Walker's seat was right next to the President's chair.

"Yes, sir." He leaned closer to Reagan. "This State dinner was not a good idea."

"I can see that indeed it wasn't."

The formal dinner didn't last long, about two hours, because all the soldiers didn't eat a bite. They tried, but it didn't work. The soldiers hadn't ever seen the food served to them in their entire lives. And the wait-staff brought the plates and the main course without any rice.

The group of soldiers went out later and found a restaurant that served rice. If they don't have rice, they don't eat.

Julius acted calm, like everything was under control. Inside he had different feelings.

"Oh, I was on pins and needles," he said to me later. "I couldn't tell what the Liberians were going to do next."

Based on my experience, I could guess what the soldiers would do, but I knew they weren't going to eat. However, the State Department doesn't listen to anybody but themselves.

The dinner was a fiasco.

"You would be better off having a ceremony for them, instead," I had recommended to the State Department.

When the dinner was complete, Doe and his soldiers came to me and talked to me as we left the building.

"I want you to go with me," Doe said.

"Where are you going?" I said.

"We're flying in a big jet to Fort Benning, Fort Bragg, and several other bases. At the end, we will fly to Disneyland."

"Bob," Julius Walker said, "Would you go with them? You've got to keep them under control."

"I'll go everywhere." I raised a finger. "But I'm not going

to Disneyland with this bunch of people. Keeping up with them would be impossible at Disneyland."

We went to Fort Bragg and several other installations. At Fort Campbell, a Brigadier General (BG) met with Doe.

"Chiefo, come here." Doe motioned me over.

I walked to his side.

"Who in the heck are you?" the BG said.

"I'm the babysitter for Sergeant Doe," I said.

He looked at me.

"Where he goes, I go." I stared back.

That ticked off the BG.

"If you should have any questions," I said, "call President Ronald Reagan."

Doe wasn't going to go anywhere without me being right next to him. We got on Air Force Two to fly to our destinations. Of course, President Doe wanted to go all around the United States on this airplane. I had to keep them in line on the plane as well as on the ground.

"Everyone sit." I pointed to the seats. "Don't walk around the plane while it's in flight. Sit tight."

They all sat.

"They're going to serve you food in a minute. But you sit right where you are and stay there."

The flight attendants brought out the food, and even though

they ate, it wasn't much. After we flew to several installations, the last stop to change planes for Disneyland was San Luis Obispo.

And I had already told Julius that was my last stop with them.

"I'm not going to Disneyland," I said to him. "Can you imagine me trying to monitor this mob in Disneyland? They'll be scattered all over the amusement park and I wouldn't be able to gather them back together."

"And I don't blame you, Bob." He shrugged. "I'll make sure the President knows you returned to duty."

The next day I caught a flight back home. I don't know what happened at Disneyland.

"We'll send a State Department man with them," Julius had said.

"That will be good training for him." I grinned. "Good training in handling foreign dignitaries and executing State Department policies in real life."

One of the great things about the Liberian visit is that Jean took it all in stride. It was one more part of being in the military. She'd been used to similar events happening throughout my career, but she made an extra effort of handling my extra commitments and duties as I worked with the Liberians, including having her vacation in Acapulco interrupted.

CHAPTER 11

CHANGE OF VENUE

After Liberia, I returned to Texas in August 1980 and completed a variety of missions working out of Fort Hood. The first assignment I had was as the III Corps Assistant Chief of Staff (ACofS), G5, which was the principal staff officer to the Commanding General for Civil Military Operations and Community Relations. In that position I supervised the approval, support and execution of approximately one hundred fifty community relations events, all with great success. I represented the Fort Hood Commander to five Chambers of Commerce, to the Central Texas Council of Government and to the Bell County

Commissioner's Court and County Rehabilitation Center. As my boss said, I was "attuned to the political sensitivities of Central Texas and [was] highly popular with the civilian population."

I enjoyed various other assignments at Fort Hood until I retired in 1984 with over thirty years of service to the Nation. My adventures didn't end with Liberia, but I'll save the rest of those stories for my family.

When I retired from the U.S. Army, I worked with businesses in Texas and I traveled with my wife Jean. We've been to Alaska, Hawaii, and Central America. I continue to enjoy my life, my family, and encouraging others to get involved in their own life adventures. I'm proud of all my children and grandchildren have accomplished.

I hope this book inspires you to lean forward in life, take advantage of the opportunities in front of you, laugh at some of the hard knocks you'll encounter, and change someone else's life for the better.

APPENDIX 1

ROBERT GOSNEY, U.S. ARMY (RETIRED)
AWARDS AND DECORATIONS

AWARD/DECORATION IN ORDER OF PRECEDENCE

- Defense Superior Service Medal
- Legion of Merit w/1 Oak Leaf Cluster
- Distinguished Flying Cross w/1 Oak Leaf Cluster
- Bronze Star Medal w/1 Oak Leaf Cluster
- Meritorious Service Medal w/1 Oak Leaf Cluster
- Air Medal w/2 V Devices w/Numerals 29
- Army Commendation Medal
- Army of Occupation Medal (Germany)
- Humanitarian Service Medal
- National Defense Service Medal
- Army Service Ribbon
- Overseas Service Ribbon w/Numeral 1
- Vietnam Service Medal w/1 Silver Service Star
- Republic of Vietnam Campaign Medal
- Master Army Aviator Badge
- Valorous Unit Award
- Vietnamese Cross of Gallantry w/Palm
- General Staff Identification Badge

APPENDIX 2

JEAN GOSNEY'S DEPARTMENT OF STATE MERITORIOUS HONOR AWARD

CITATION

THELMA JEAN GOSNEY

For consistent, dedicated and superlative service as a consulate assistant, for initiative in developing an anti-fraud system, for service above and beyond the call of duty in areas outside her immediate responsibility, and for steady, gracious manner which made all clients feel welcome and well served.

Signed: **Robert P Smith, Ambassador**

This award was presented to Jean Gosney upon her departure from Liberia with her husband Colonel Bob Gosney, for service involved during and after the coup of 12 April 1980 in Liberia.

Department of State

United States of America

Meritorious Honor Award

THELMA JEAN GOSNEY

For consistent, dedicated and superlative service as a consulate assistant, for initiative in developing an anti-fraud system, for service above and beyond the call of duty in areas outside her immediate responsibility, and for steady, gracious manner which made all clients feel welcome and well served.

August 1980

Ambassador

ABOUT THE AUTHOR

ROBERT GOSNEY

COLONEL, U.S. ARMY (RETIRED)

Robert R. Gosney is a native Texan and retired U.S. Army Colonel. He's served in Europe, Korea, Vietnam, Liberia, and around the United States. His life has impacted tens of thousand of individuals, particularly in Vietnam and Liberia. He's placed himself in extreme danger regardless of personal safety. He's received numerous military awards for service and valor in battle. He and his wife, Jean, reside in Belton, Texas, and have four children, and many grand- and great-grandchildren.